The Eureka! Moment

The Eureka! Moment

100 Key Scientific Discoveries of the 20th Century

Rupert Lee

THE BRITISH LIBRARY

First published 2002 by
The British Library
96 Euston Road
London NW1 2DB

British Library Cataloguing-in-Publication Data
A catalogue record for this book is available from The British Library

ISBN 0–7123–0884–9

Designed by Andrew Shoolbred
Typeset by Hope Services (Abingdon) Ltd
Printed in England by Biddles, Guildford

Author's Preface and Acknowledgements

In 2001 The British Library commissioned this book, describing a hundred of the twentieth century's major scientific discoveries. When I talked about the project to non-scientists, I soon found that most people associate science with inventions, and have little idea of the discoveries that lie behind them. A typical comment was 'I trust you will include the mobile phone in your list'. Of course, the mobile phone is an invention, not a discovery. However, the dividing line between discoveries and inventions is not always clear-cut, and there were some borderline cases which were hard to decide on. For instance, the Haber Process (a cheap method of making ammonia out of nitrogen and hydrogen) can equally be described as a discovery and as an invention. In this case I decided to include it, because it has had a major impact on the chemicals industry. However, as another example, I decided that the Polymerase Chain Reaction (a key technique in genetic engineering) was really an invention and not a discovery at all, so it has been left out. Another borderline case was the transistor. In this case I decided to include an essay describing the underlying science, but mentioning only briefly the difficulties its originators had in putting it into practice.

I am aware that some scientists may not approve of this book's portrayal of science. They will say, rightly, that major 'eureka moments' are by and large not what drives science forward. Most scientific progress is the cumulative effect of lots of smaller discoveries, often just as hard-won as the more spectacular breakthroughs. I am reminded of a cartoon I once saw, showing a record shop with two racks, one labelled 'Opera Highlights' and the other 'Opera Draggy Bits'. Charting the history of twentieth-century science simply by describing some of its biggest discoveries is rather like trying to appreciate a work of classical music by listening only to its louder climactic moments. However, nobody can deny that some discoveries are a lot more important than others, so a book describing some of the most important ones should have its place. The phrase 'eureka moment' also seems to imply a single blinding flash of inspiration, while in reality most of these discoveries were much longer drawn-out affairs. However, I contend that in every discovery there comes a particular day when the scientists involved (there are usually more than one) are able to raise a cheer and say, 'Now we've got it'. It is these moments that make all the hard slog of scientific research bearable.

This book draws entirely on material held in the collections of The British Library. Each essay provides a citation to the document (usually an article in a learned journal) where the discovery was first published. All but a few of these are held by The British Library, and information on the few that it does not hold was found in review articles, monographs, and biographies in The British Library's collections. All of this material can be read in the Reading Rooms of the Library's St Pancras building in Euston Road, London, and most of it can also be obtained by other libraries on inter-library loan from The British Library Document Supply Centre at Wetherby, West Yorkshire.

I would like to thank Anthony Warshaw and his colleagues in The British Library Publications Office for commissioning this book, and for all their help in bringing it to completion. I am also grateful to the following organisations for providing illustrations and granting us permission to reproduce them in whole or in part: Science Museum/Science & Society Picture Library, London (plates 1, 2, 4-7, 9, 12 and 13), the Jewish National and University Library, Jerusalem (plate 3) and The British Library Board (plate 14). Finally I would like to thank all the various professional scientists who kindly suggested discoveries that I might include, and commented on different parts of the original draft. They suggested several improvements, and also spotted a few factual errors which I was pleased to rectify. They are not in any way responsible for any inaccuracies that may remain.

Rupert Lee
The British Library
July 2002

Contents

LIST OF PLATES xv

INTRODUCTION 1

MEDICINE 14

The Yellow Fever Virus (Reed, 1900) 18
Blood Groups (Landsteiner, 1901) 20
Vitamins (Hopkins, 1912) 22
Insulin (Banting and Best, 1922) 24
Brain Waves (Berger, 1929) 26
Penicillin (Fleming, 1929) 28
Sulphonamides (Domagk, 1935) 30
Rejection of Transplants (Gibson and Medawar, 1943) 32
Sickle Cell Anaemia (Pauling et al, 1949) 34
Smoking Causes Lung Cancer (Doll and Hill, 1950; Wynder and Graham, 1950) 36
The Polio Vaccine (Salk, 1953) 38
Interferons (Isaacs and Lindemann, 1957; Isaacs et al, 1957) 40
Prions (Prusiner et al, 1983) 42
HIV (Barre-Sinoussi et al, 1983) 44
Helicobacter pylori (Marshall et al, 1985) 46

Embryonic Stem Cells Grown Artificially (Gearhart, 1997) 48

HISTORY OF LIFE 50

First Clues About How Life Began (Miller, 1953) 54
The Early Presence of Life on Earth (Mojzsis et al, 1996) 56
The Origin of Higher Organisms (Sagan, 1967) 58
Nemesis of the Dinosaurs (Alvarez et al, 1980) 60
'Lucy' (Johanson and Taieb, 1976) 62
The 'Taung Child' (Dart, 1925) 64

BIOLOGY 66

Conditioned Reflexes (Pavlov, 1906) 70
Bacteriophages (d'Herelle, 1917) 72
The Vertebrate Organiser (Spemann and Mangold, 1924) 74
Urease (Sumner, 1926) 76
Progesterone (Corner and Allen, 1929) 78
The Krebs Cycle (Krebs and Johnson, 1937) 80
ATP (Lipmann, 1941) 82
Bee Dances (von Frisch, 1946) 84
Evolution Driven by Competition (Lack, 1947) 86
How Nerves Work (Hodgkin et al, 1952; Hodgkin and Huxley, 1952) 88
The Chemical Composition of Insulin (Sanger and Thompson, 1953) 90
The Structure of Vitamin B12 (Hodgkin et al, 1956) 92
The Calvin Cycle in Photosynthesis (Calvin, 1962) 94
Monoclonal Antibodies (Kohler and Milstein, 1975) 96

GENETICS 98

One Gene—One Enzyme (Garrod, 1909) 102
Genes on Chromosomes (Morgan, 1910) 104
Genes are Made of DNA (Avery et al, 1944) 106
Gene Recombination in Bacteria (Lederberg and Tatum, 1946) 108
Transposons (McClintock, 1951) 110
The DNA Double Helix (Watson and Crick, 1953) 112
The Central Dogma (Crick, 1958) 114
The Nature of the DNA Code (Crick et al, 1961) 116
The Operon (Jacob and Monod, 1961) 118
Reverse Transcriptase (Baltimore, 1970; Temin and Mitsuzani, 1970) 120
Oncogenes (Martin, 1970) 122
Sequencing a Genome (Sanger et al, 1977) 124
The Antibody Problem (Tonegawa, 1983) 126
Homeobox Genes (McGinnis et al, 1984) 128
DNA Fingerprints (Jeffreys et al, 1985) 130
'Dolly', the Cloned Sheep (Wilmut et al, 1997) 132

ASTRONOMY AND COSMOLOGY 134

Galaxies (Hubble, 1925) 138
The Expanding Universe (Hubble, 1929) 140
Pluto (Tombaugh, 1930) 142
What Makes Stars Shine (Bethe and Critchfield, 1938; Bethe, 1939) 144
The Age of the Solar System (Patterson, 1955) 146
The Origin of the Chemical Elements (Burbidge et al, 1957) 148
Quasars (Hazard et al, 1963; Schmidt, 1963; Oke, 1963; Greenstein and Matthews, 1963) 150

Proof of the Big Bang (Penzias and Wilson, 1965) 152
Pulsars (Hewish et al, 1968) 154
Organic Molecules in Interstellar Space (Snyder et al, 1969) 156
Gamma-Ray Bursters (Djorgovski et al, 1997; Metzger et al, 1997) 158
Planets Orbiting Other Stars (Mayor and Queroz, 1997) 160
The Age of the Universe (Freedman et al, 2000) 162

EARTH SCIENCES 164

The Earth's Core (Oldham, 1906) 168
The Moho (Mohorovicic, 1909) 170
The Ionosphere (Appleton and Barnett, 1925) 172
Plate Tectonics (Morgan, 1968) 174
Methane Clathrate in the Ocean Depths (Hollister et al, 1972) 176
Black Smokers (Spiess et al, 1980) 178
The Ozone Hole (Farman et al, 1985) 180

CHEMISTRY 182

The Haber Process (Haber, 1909) 186
Chemical Bonds (Lewis, 1916) 188
Polymers (Staudinger and Fritschi, 1922) 190
Plutonium (McMillan and Abelson, 1940) 192
Ziegler-Natta Catalysts (Ziegler et al, 1955) 194
Buckminsterfullerene (Kroto et al, 1985) 196
High-Temperature Superconductors (Muller and Bednorz, 1986) 198

PHYSICS 200

The Quantum Theory (Planck, 1900) 204
Radioactive Decay (Rutherford and Soddy, 1903) 206
Proof That Atoms Exist (Einstein, 1905) 208

Photons (Einstein, 1905) 210
Special Relativity (Einstein, 1905) 212
Superconductivity (Kammelingh Onnes, 1911) 214
The Atomic Nucleus (Rutherford, 1911) 216
Cosmic Rays (Hess, 1912) 218
General Relativity (Einstein, 1915) 220
The Proton (Rutherford, 1919) 222
The Quantum Atom (Bohr, 1923) 224
The Uncertainty Principle (Heisenberg, 1927) 226
The Positron (Anderson, 1932) 228
The Neutron (Chadwick, 1932) 230
Nuclear Fusion (Rutherford et al, 1933) 232
Nuclear Fission (Hahn and Strassmann, 1939) 234
The Transistor (Bardeen and Brittain, 1948) 236
The Neutrino (Reines and Cowan, 1953) 238
The Hunting of the Quarks (Friedman, 1972) 240
The W and Z Particles (Arnison et al, 1983) 242
Bose-Einstein Condensate (Anderson et al, 1995) 244

SCIENCE IN THE YEAR 2000 246

INDEX 259

List of Plates

1. Einstein at the blackboard in the 1930s. Around this time he was trying unsuccessfully to disprove the theories of Niels Bohr and Werner Heisenberg (*see* The Quantum Atom and The Uncertainty Principle).

2. Rutherford in the Cavendish Laboratory, Cambridge, in the early 1930s: talking to a colleague, and dropping cigar ash around sensitive equipment. His voice could probably be heard through the wall. Parts of this building have since been sealed off behind lead and concrete because of the dangerous levels of radioactive contamination.

3. 'Das Gesetz von der Aequivalenz von Masse und Energie ($E = mc^2$)'—the Law of Equivalence of Mass and Energy. During the course of his career, Einstein found several different ways of deriving his famous equation. This is a late manuscript, dating from 1946 (*see* Special Relativity).

4. Physics then. Ernest Rutherford's laboratory bench at McGill University, Canada, in the early 1900s. McGill's physics laboratory was then one of the best equipped in the world. It was here that Rutherford and Frederick Soddy showed that radioactivity was caused by unstable atoms disintegrating spontaneously (*see* Radioactive Decay).

5. Physics now. One of the four particle detectors attached to the Large Electron Positron collider (LEP) at the CERN Laboratory, Geneva, Switzerland. At the end of 2000 the LEP was dismantled to make way for an even bigger collider (*see* Science in the Year 2000: Physics).

6. The Great Nebula in Andromeda. The bright centre of the nebula is just visible to the naked eye as a dim fuzzy patch of light, appearing rather smaller than the Moon. This time-exposure shows the nebula's full size. In the 1920s Edwin Hubble showed that it is an entire separate galaxy, over a million light-years away from our own (*see* Galaxies).

7. Astronaut Harrison Schmidt, the only professional geologist to have visited the Moon, on the Apollo 17 mission in 1972. Rocks collected by Schmidt helped to confirm the age of the Solar System as 4.55 billion years (*see* The Age of the Solar System).

8. Plate tectonics. The original map published by W. H. Morgan in 1968, showing the different geological 'plates' superimposed on a map of the continents. Over periods of tens or hundreds of millions of years, the positions of the continents shift due to movements of the plates. These movements cause earthquakes and volcanic eruptions around the plates' edges.

9. The discovery of Pluto. These two photographs of stars were taken three days apart by Clyde Tombaugh at the Lowell Observatory, Flagstaff, Arizona in March 1930. One small 'star' (*marked with arrows*) has moved. It must be a planet.

10. 'Lucy', the skeleton of a female *Australopithecus afarensis* discovered by Donald Johanson in 1975. This is the most complete skeleton of a fossil hominid to have been found so far, and at four million years it is one of the oldest. Lucy

stood fully upright like a modern human, and was probably rather less than 1.2 metres tall. (Males of this species were bigger.)

11. Diagram of the DNA molecule, shown here reproducing itself. The two strands unwind from each other, and a new strand then builds itself along each of the original strands, resulting in two identical copies of the original molecule. The 'letters' of the genetic code are the 'bases' A, C, G and T. As the two new strands are built, C bases always link with Gs, and As always link with Ts.

12. DNA. James Watson with the model of the DNA molecule that he and Francis Crick had built at Cambridge in 1953. At the time when this photograph was taken in 1994, Watson was Head of the Human Genome Project.

13. The discovery of penicillin, 1928. Colonies of the bacterium *Staphylococcus aureus* (small round blobs) are being grown on a layer of gelatin in a glass plate in Alexander Fleming's laboratory. The plate has been accidentally contaminated by the mould *Penicillium notatum.* The *Staphylococcus* colonies nearest the *Penicillium* are being killed off by a chemical secreted by the mould.

14. The fruit fly *Drosophila melanogaster*, female and male. This fly was first used for genetics research by Thomas Morgan, who used it in 1910 to show that an organism's genes are located on the chromosomes that are found in every cell of its body. It has since become one of the most intensively studied of all organisms, and has provided us with much of our present knowledge of genetics (*see* Genes on Chromosomes; Homeobox Genes).

15. Bee dances. Foraging honey-bees perform two different kinds of dance on the surface of the honeycomb to show

their co-workers where to find flowers with nectar. The 'round dance' means 'there are flowers near the hive'. The 'waggle dance' shows the direction of more distant flowers. The angle from the vertical at which the dance is performed shows the direction of the flowers relative to the Sun.

INTRODUCTION

Posterity will remember the twentieth century for many things. The history books of the future will inevitably dwell on such social and political events as the two World Wars; the rise and fall of Marxist-Leninism; the arrival of the United States of America as a global superpower; the end of the British Empire and of European colonialism generally; and, quite possibly, the beginning of the rise of China and the European Union as economic powers of global importance. But above all, it must surely be remembered as a time of unprecedented growth in the economies of the industrialised nations. Despite the two most destructive wars ever fought in the whole of history, a period of severe economic depression in the 1920s and 1930s, and numerous lesser economic downturns, the self-styled 'First World' ended the century incomparably more wealthy than it had begun it. The causes of this extraordinary rise in prosperity will doubtless be analysed at length by historians, but one factor is certain to stand out: the gathering momentum of science and technology. More technology resulted in the creation of more wealth, which financed more science, which gave rise to even more technology, creating yet more wealth, and so on.

It may seem strange to us today, to realise that people have not always appreciated the link between science and technology. Until the middle of the nineteenth century, science was largely a scholarly pursuit for gentlemen amateurs, while the engineers who brought about the Industrial Revolution and the 'Age of Steam' often had little scientific knowledge, or indeed education of any kind. For instance George Stevenson, builder of the Manchester-Liverpool railway and the famous 'Rocket' locomotive of 1829, was well advanced in his career before he even learnt to read and write. The link between scientific research and industrial innovation seems to have first been established in the chemical industry, with the inventions of celluloid and mauve dye in the 1850s perhaps marking the birth of industrial science as we now know it. The next few decades also saw the arrival of the electrical industry, the first to have come into being as a direct result of discoveries made by professional scientists. Chemistry and electricity between them spawned a second industrial revolution in the late nineteenth century, a fact which receives little mention in British history textbooks, probably because it was America and Germany that made the running. Companies like General Electric and I.G. Farben quickly became industrial giants, while British technology, which had led the first Industrial Revolution, was left trailing far behind. This was the period in which modern scientific culture may be said to have its roots. As the technology-based industries grew, so did the need for qualified professional scientists, and in Europe and North America there was a proliferation of university science departments to meet this demand. From the start, many (but not all) of these were involved in active research as well as teaching.

So the twentieth-century burgeoning of science really began during the second half of the nineteenth. However, science in the twentieth century had one notable feature all of its own, namely the so-called 'big science' programmes. Previously, science had been a rather lonely pursuit, with each researcher working on his (very rarely her) own, with at most a handful of semi-qualified assistants. Science in the twentieth century, by contrast, became dominated by teamwork, with dozens or even hundreds of researchers working together on large-scale projects, commanding large budgets. 'Big science' began in the century's first decade. Perhaps the first big team was that set up by Ernest Rutherford, when he became Professor of Physics at Manchester University in 1907, his head full of ideas for research on the structure of the atom. Realising that he could not begin to do all this research on his own, he recruited a number of graduate students and set them to work together. Over the next forty years, such teamwork increasingly became the norm in experimental physics and chemistry. Biology and medicine were slower to follow suit, and indeed a few one-person research projects still go on in zoology and botany departments to this day.

The advantages of such teamwork are obvious. Experiments can go on round the clock, and most importantly, team members can each contribute their own expertise, and can bounce ideas off each other. A classic case was the race to discover the structure of the DNA molecule in 1953. At Cambridge, Francis Crick and James Watson collaborated, while at King's College, London, Rosalind Franklin tried to solve the problem largely on her own. Since Franklin had much more data to work on than Crick and Watson, she should have been able to find the answer first. Instead, she lost valuable time pursuing false trails, and was eventually beaten to it. Nobody has suggested she was a less clever scientist than Crick and Watson; she is generally acknowledged to have been a brilliant researcher. Her trouble was that she did not discuss her work with her colleagues, indeed she was barely on speaking terms with some of them. A more frightening example of the benefits of teamwork was the Manhattan Project. During the Second World War, both the United States and Germany decided to research into the feasibility of nuclear weapons. The Americans quickly set up a huge team project, which in due course produced two different designs for atomic bombs, both of which were dropped on Japan. In Germany, meanwhile, the Nazi government left the physicists to work on their own in separate laboratories, and they never even solved the basic theoretical problems. Had the Germans brought their nuclear physicists together into a single team, the course of history might have been very different.

However, 'big science' has its drawbacks. Large teams can often be beset with bureaucracy. They may have to set their goals years in advance, and find it hard to change direction at short notice. The problem is most acute if non-scientists set their goals for them. This seems to have been one of the troubles with science in the Soviet Union. The Communist government always valued science very highly, and established a number of large research institutes. Scientifically-literate people were found at high levels in their political and civil service hierarchy (unlike in the Western nations, where the ruling elites consisted almost entirely of people educated in the arts and humanities, who often seemed to take a positive pride in their ignorance of scientific matters). The Soviet Union also had several brilliant scientists who won international respect. And yet their achievements generally lagged behind the West. Not one of the discoveries described in this book came out of the USSR. (Pavlov's discovery of the conditioned reflex was made before the Russian revolution.) Doubtless there were several reasons for this underperformance, but the way in which science was administered was very likely a contributing factor. It seems that the big institutes did not sufficiently encourage original thinking by individual researchers.

The research institutes in the Soviet Union and the Manhattan Project in the United States were two examples of another major phenomenon in twentieth-century science: research run by governments. This was not an entirely new development: after all, the British government had been running the Royal Greenwich Observatory ever since the seventeenth century. But around the early 1900s governments started to realise that just as science brought profits to industry, so too it could have a pay-off in terms of political power. It can well be argued that the First World War was made possible by Fritz Haber's discovery of a process for making cheap ammonia, the raw material from which both synthetic fertilisers and explosives are made. The fertilisers made agriculture more productive and less labour-intensive, enabling governments to recruit and maintain larger armies than ever before. The cheap explosives then made possible the immense artillery bombardments of the Western Front. The impact of Haber's science was not lost on governments, who were soon running research programmes in chemistry, radio, aeronautics, medicine (in Britain, the Medical Research Council was set up during the First World War) and, in due course, nuclear energy. To start with, these were mostly military projects. However as the century progressed, and government research programmes expanded, they increasingly spilled over into civilian fields. After World War Two, the victorious nations started massive government-run research projects in civilian uses for nuclear power. This was also the era of the 'space race' between the United States and the Soviet Union (later joined by other nations on a smaller scale). The rocket technology was of military origin, and the race to reach the Moon and achieve other space 'firsts' was at least partly motivated by defence considerations, but much of the American and Russian space programmes were (and still are) devoted to entirely peaceful scientific research.

The first half of the century saw a huge expansion in the fields of physics and chemistry. The second half saw medicine and biology start to catch up. Medical research, unlike the other sciences, had been a job for professionals throughout history, but like other sciences it had received hardly any funding until the second half of the nineteenth century, and progress had been very slow and haphazard. In the twentieth century, the rate of progress increased hugely, and 'big science' finally reached the life sciences in the 1930s, when a team was set up at Oxford University under the leadership of Professor Howard Florey, to find a way of turning the recently-discovered penicillin into a workable drug. After the Second World War large research teams increasingly became the norm in biology, as they were in physics and chemistry.

The penicillin project was funded by the British government, but by then biochemistry was also becoming firmly commercialised, as the pharmaceutical industry increasingly invested in serious scientific research. At the beginning of the twentieth century, pharmaceutical manufacturers were still small- to medium-sized businesses, producing 'patent medicines' and other nostrums, often with a strong whiff of quackery. The invention of Salvarsan, a drug to treat syphilis, in 1909, may mark the point at which pharmaceutical manufacturers realised the value of biochemical research. They quickly started setting up their own laboratories, and the success of their new science-based products led to a huge expansion of the industry. In the closing years of the century, a series of mergers resulted in a small number of pharmaceutical laboratories coming to dominate medical research—to many peoples' unease. In the immediate post-war years, the Western democracies and the communist Eastern Bloc nations had all set up 'welfare states' or similar social provisions, in which it was understood that medicine and health-care were to be provided as a service for the public good. Many people feared that the rise of the pharmaceutical giants meant that this principle was being eroded, and that medicine was again becoming a business pursued mainly for profit, with the inevitable result that poor people would not be able to afford it. The problem was particularly acute in the developing nations. As the century ended, the long-hoped-for spread of First-World standards of health care to developing countries seemed not to be happening, as modern medicine priced itself out of the Third World market. Finding ways out of this dilemma became a serious political problem, debated at length by scientists themselves as much as by everyone else.

The influence of the pharmaceutical companies is only one of many areas where science has provoked controversy. As science has impinged more and more on society, so the debate has intensified over who should make the decisions over what research the scientists pursue, and how their discoveries should be used. The scientists themselves have generally argued that they should be given maximum freedom to set their own targets. Scientists are driven by a wide variety of motivations, but they all have high levels of curiosity and inquisitiveness. Generally, wherever there is a question to be asked about the workings of nature, and even a hint of a way to find the answer, there will be scientists eager to get to work on it. They will not much like being told by non-scientists which research topics to pursue and which to leave. However, as research methods become more sophisticated, they become increasingly expensive, and the funding comes in one way or another from society at large. The general public does not necessarily share the scientists' enthusiasms, and needs to be convinced that the money is being spent wisely. As the twentieth century progressed, scientists increasingly found they had to justify their research to a sceptical and sometimes even hostile public. Why, they were asked, should the world spend large sums on research into fields like cosmology, that are unlikely to yield practical benefits for a long time, if ever? Do scientists, immured in their laboratories and obsessed with their pet projects, appreciate the social and ethical implications of their research in fields such as molecular biology or nuclear energy? How can scientists, who flourish best in times of peace and prosperity, justify research on military projects?

These are serious issues. It is not just a question of preventing the kind of pseudo-science that was practised by doctors and psychiatrists in Nazi Germany and Stalinist Russia. There have been many instances of ethically compromised science in the free world as well. One need only think of the widespread practice of frontal lobotomy in America after the Second World War, as a way of 'curing' shell-shock and post-traumatic stress by turning the sufferer into an unresponsive zombie; or the way in which the British and American armies deliberately exposed personnel to nuclear bomb tests in the 1950s in order to study the effects of radiation. Science policy is not necessarily best left to the scientists themselves. Military research, of course, will always have strong opponents among people with pacifist inclinations—including many scientists. However, in the latter half of the twentieth century there was also growing unease about peaceful research, and indeed about the whole of science generally. In particular, advances in medicine and biology had profound social and ethical implications, which the scientists themselves sometimes seemed slow to appreciate. The anti-vivisection movement, opposition by the Roman Catholic Church to the contraceptive pill, and (in Britain) the public backlash against genetically modified foods, were all instances where non-scientific society felt that scientists had failed to understand the moral dimension to their research. In the case of the Pill, this was just part of a much wider debate between social 'progressives' and 'conservatives', but on several other issues, such as vivisection, the dividing line was more clearly drawn between pro- and anti-science factions.

In these debates the problem was often made worse by a lack of communication. To put it bluntly, scientists are often bad at explaining themselves. In the 1960s C. P. Snow famously decried the 'two cultures', by which our education system forced people to specialise in either the sciences or in the humanities from their mid-teens onwards, and people on each side of the divide grew up with little understanding or respect for the other side. Since then, the only real progress in bridging this divide has been the rise of 'popular science', in books, newspaper articles and television programmes, in which a few scientists try to explain their science to the wider public (often making themselves unpopular with their peers in the process). The genre has been highly successful, with some books becoming bestsellers, and television programmes often getting very high ratings. However, scientists are still faced with the problem of explaining their work to people who find the entire subject repugnant. A whole strand of anti-science thinking, which might perhaps be summed up by the single word 'Frankenstein', runs through Western culture. The potency of this sentiment was understood by H.G. Wells, whose movie *Things To Come*, made in the 1930s, showed a mob of anti-science protesters trying to halt the launching of the first astronauts to the Moon. His prophecy seemed apt in the summer of 1969, when Armstrong and Aldrin actually landed on the Moon just as the hippie counter-culture was at its height. Towards the end of the century, some anti-science protesters really did go on the warpath, with self-styled 'eco-warriors' and 'animal rights campaigners' trying to sabotage research projects, and even murder scientists. Scientists often complain that such anti-science sentiment is a result of ignorance, but if this is so, then they themselves should perhaps accept some responsibility for failing to put their viewpoint across.

The rise of science in the twentieth century, and its backlash, have been very largely a phenomenon of the Western industrialised democracies. Other parts of the world, notably Japan, the Eastern Bloc, China, and even some Third-World nations, also have (or had) science programmes, but they have all built their research on a foundation of Western First-World science. Some will try to argue that this is because Western culture has been uniquely conducive to science, especially since the Renaissance and the Enlightenment period of the early eighteenth century. However, it is worth noting that until five hundred years ago, it was the Muslim world that was pre-eminent in science, while Christian Europe was mired in the Middle Ages. Nevertheless, the science of the twentieth century has its roots in those cultures which, in the centuries immediately preceding, most encouraged open intellectual enquiry: Protestant Christian, liberal Jewish, and humanist. A hundred years ago, this meant that the world's centre of science was Germany, Switzerland and Austria, with Britain and France (nominally Roman Catholic, but with a strong tradition of atheist humanism) vying for second place. German came near to being the international language of science. Between 1900 and 1932, just one hundred Nobel Prizes were awarded in the sciences: thirty-three of these went to German and Swiss citizens, eight of whom were Jews.

Then came Hitler. The Jewish scientists fled continental Europe, mostly to Britain and the USA, whose universities generally made them welcome. (An exception was Fritz Haber, who was regarded with distaste because of his role in the development of chemical weapons.) They took to publishing their research in English-language journals. As a result, science's 'centre of gravity' moved to Britain and North America, and English emerged as the international language for scientific writing. Some other countries, notably Russia and China, still encourage their scientists to publish their research in their native languages, but they are fighting a losing battle. A look at the languages of the publications described in this book shows this clearly. Of the thirty-five discoveries dated between 1900 and 1932, fifteen were published in German, nineteen in English, and one in French. The sixty-five discoveries dated from 1933 onwards were all published in English apart from five, which were published in German.

This puts scientists who do not speak fluent English at a disadvantage. We have seen how scientists benefit from being able to discuss their ideas with each other, and in these discussions non-English-speaking scientists clearly find themselves 'out of the loop'. An obvious instance was Nicholas Paulesco, who might have shared in a Nobel Prize for the discovery of insulin, along with Banting and Macleod. His research was almost completely overlooked, simply because he published it in French. Living in Romania, and unable to speak English, he could not collaborate or exchange ideas with other scientists in his field, and although he started well ahead of Banting's team, he was soon left behind. Another example is the discovery of buckminsterfullerene in 1986 by British and American chemists. They were quite unaware that this molecule's existence had been predicted some fourteen years earlier, in a journal article written in Japanese. Japan has a massive science programme, which is to a large extent cut off from the rest of the world by the language barrier. Russian and Chinese scientists suffer a similar disadvantage. Undoubtedly one of the reasons why Soviet science lagged behind the West was a communication problem, compounded of the Communists' culture of secrecy and their insistence on writing in Russian. So whether we like it or not, a single language for all scientists is the most sensible arrangement, and it seems that English is well on its way to being that language.

The twentieth century has truly been the century of science. In 1900, scientists were still a small, marginal group in society, often applauded for their achievements but equally often smiled at for their eccentric pursuits. In Britain, for instance, it is reckoned that there were only about 2400 professional scientists all told. A hundred years later, they were part of society's mainstream, regarded with respect sometimes verging on awe, but often also with suspicion and resentment. They had specialist knowledge, unintelligible to outsiders, that brought them (or their employers) great power and wealth. The rich nations and big companies which could afford large-scale research programmes reaped the benefits and became richer still, while the poorer nations and smaller companies tended to lose out. This is not to deny that there have been numerous instances where scientific advances have helped the lot of the world's poor. The 1960s' 'green revolution' in Third-Word (especially Asian) agriculture is a clear example. Nevertheless, modern technology and the science that lies behind it are one of the main factors (possibly the main factor) involved in the widening gap between the rich and poor that was such a clear trend of the late twentieth century. It is ironic that scientists, whose political views have on average tended to be rather to the left of their non-scientific contemporaries', have contributed to social inequality in this way. There is no knowing whether this trend will continue for long in the twenty-first century, but there are at present few signs of it ending.

MEDICINE

Of all the sciences, medicine surely has the most direct and immediate impact on people's lives. One only has to contrast a visit to the doctor or hospital in the 1990s with a similar visit in the 1890s to appreciate the achievements of science during the century. A hundred years ago, if you went to the doctor (or, just as often, he came to see you), he would often be able to make some diagnosis, but the number of treatments he could offer were very few. Pharmacists could offer a variety of tonics, purges, patent medicines and nostrums, some of which may have been helpful on occasions, but the chief weapon in a doctor's armoury was really nothing more than a reassuring bedside manner. Surgeons, meanwhile, could claim to be rather more advanced. The discovery of bacteria, and the arrival of antiseptics to reduce the risk of infection, had already led to major reductions in the mortality rates among their patients. Even so, a simple operation such as having one's appendix removed was still highly risky. The advances of the next hundred years were to change the practice of medicine almost beyond recognition.

The twentieth century saw immense improvements in peoples' overall levels of health, especially in the Western world. With each succeeding decade, average lifespans have continued to increase. Most people, if asked, would doubtless say that much of this progress resulted from the discoveries made by medical science during the century. They might point out that pneumonia was a frequent killer until the arrival of sulphonamides and antibiotics; vaccinations have removed the threats from diphtheria and polio; blood transfusions reduce the risk of dying during a surgical operation; and diabetes can now be treated so that sufferers live to reasonable old age. But in truth, the biggest improvements in peoples' health have resulted not from scientific discoveries, important though these are. Public-health initiatives aimed at improving diet, hygiene and sanitation have played a much larger part. Good drinking-water supplies and proper drains have done far more to fight the spread of bacterial diseases than antibiotics ever will. The decline of alcoholism, which in the nineteenth century had been a massive health problem among the working classes, mainly resulted from the decision by Western governments to impose tax on distilled spirits, and so price them beyond the reach of poorer people. Also the arrival of freely-available health care has meant that poor people have access to medical facilities as much as the more prosperous sectors of society. Most importantly of all, improved education means that fewer people pursue unhealthy life-styles out of simple ignorance. The discovery of the link between smoking and lung cancer, for example, has had a major impact on peoples' tobacco consumption, but this is most marked in the more educated sectors of society. All these advances in public health owe as much to the large-scale application of common sense as to cutting-edge science.

Nonetheless, twentieth-century medical science still has an impressive catalogue of successes to its credit, and one might expect that it would be respected by everyone. Instead, the closing decades of the twentieth century saw a rising scepticism on the part of the general public, coupled with the increasing popularity of various 'alternative' and 'complementary' therapies, whose practitioners often openly boast of their opposition to conventional science. Much of this popular unease about medical science may result from the way it presents itself. Conventional scientists are often thought of as arrogant people who think they know all the answers, and who treat patients simply as machines that need mending. This image is a caricature of course, but it is not completely unfounded. Alternative and complementary therapists, by contrast, talk about 'healing' rather than 'curing', claim to 'treat the patient, not the disease', and generally take pains to present a more sympathetic image. Like the conventional doctors of a century ago, they know the power of a good bedside manner. However, it is noticeable that people usually turn to conventional medicine to treat anything that is truly life-threatening. Many people have reservations about it, but very few reject it outright.

Public suspicion of medical science may also have another cause. Medical research has an ethical dimension, which may sometimes be appreciated more by the non-scientific public than by the scientists themselves. The anti-vivisection movement has a strong public appeal, and society at large does not always share scientists' enthusiasm for research involving human embryos. Scientists usually respond to criticism of their ethics by asserting that these experiments are a necessary price to pay for progress. For example, the discovery of insulin entailed some decidedly cruel experiments on dogs, and it almost certainly could not have been achieved by more humane means. It has saved the lives of countless diabetes sufferers, and for most scientists this is justification enough. However, most scientists now realise that the issue is a sensitive one, and acknowledge that they must set themselves clear standards, and then abide by them carefully, if they are to retain the public's support.

In 1900, medical science was simply concerned with finding first the causes of disease, and then looking for cures. In 2000, this was still its main priority, but knowledge had advanced to the point where medicine could also set itself more controversial goals. A few scientists were openly aiming to clone humans—and inviting trouble by doing so. Most of their colleagues were convinced that this was still impractical, but few would feel at all sure that it will never be done. Medical research, which a century ago seemed unquestionably a force for good, now presents a much more ambiguous image.

THE YELLOW FEVER VIRUS

The first human virus identified

One of the greatest scientific achievements of the late nineteenth century was the discovery of germs—living organisms too small to be seen by any but the most powerful microscopes, which are the main cause of infectious diseases. In the 1880s Louis Pasteur discovered bacteria, the cause of sepsis in wounds, tuberculosis, and many other diseases. Then in the 1890s it became clear that some diseases in plants were caused by agents too small to be seen under any microscope, but which nonetheless seemed to be alive. The name 'virus' was coined for them. But it was not until 1900 that viruses were also shown to be agents of disease in animals and humans.

The first human disease that was shown to be caused by a virus was yellow fever. This disease, which was often fatal, was endemic throughout much of tropical Africa and America, extending as far north as the Caribbean. Combating it became a high priority of the US Army Medical Corps, as the USA sought to increase its presence in Latin America. In 1898 Major Walter Reed, a surgeon with experience of both laboratory research and front-line duty, was given the task of identifying how yellow fever was spread. He set up a commission, together with the bacteriologist Jesse Lazear and the physician James Carroll. One possible agent of infection that came under their suspicion was the mosquito *Aedes egypti*, which is very numerous in subtropical regions. As a preliminary test, in August 1900 Carroll and Lazear deliberately let mosquitoes bite them. Sure enough, twelve days after being bitten, they fell ill with yellow fever. Carroll recovered, but Lazear died. On the strength of this result, Reed set up a special camp in Florida, where volunteers were exposed to the mosquito. (Apparently they all survived.) The results of his experiments fully confirmed his suspicion. The blood of a yellow fever patient contained an infectious agent, and any mosquito that bit a patient would then carry the disease to its next victim.

However, searches for a bacterium in the patients' blood drew a complete blank. Reed noted that recent research on foot-and-mouth disease in farm animals showed that it too was caused by an invisible agent, that could pass through the very finest filters. He rightly suggested that yellow fever and foot-and-mouth disease were both caused by a new class of disease organisms, that were very much smaller than even the smallest bacteria. He did not himself use the word 'virus', but this soon became the accepted term.

On the strength of Reed's discovery, the US Army sent the surgeon William Gorgas to Havana, Cuba, with the task of eradicating the mosquito *A. egypti*. He was so successful that within a year yellow fever had virtually disappeared from that city. He then repeated the operation in Panama, enabling American contractors to start building the Panama Canal through what had previously been one of the most fever-ridden regions in the world. However, Reed himself was unable to play any further part in the story, having died of appendicitis in 1902.

W. Reed, *Philadelphia Medical Journal*, 6, 790 (1900)

BLOOD GROUPS

Blood transfusions made possible

Blood transfusion is not a new practice. The Incas in Peru were doing it in the sixteenth century when they were discovered by the Spanish conquistadors. However, when European doctors in the seventeenth century attempted to infuse their patients' blood with blood taken from other people, the results were often disastrous, resulting in severe fever and even death. It seemed that one person's blood was not always compatible with another's. If one tried to mix them, they often congealed into a sticky mass. The practice was dropped after a few experiments.

However, the idea was not entirely forgotten. The early immunologists of the late nineteenth century were soon asking: why does blood transfusion work in some cases, but not in others? This was the problem taken up by Karl Landsteiner, an assistant at the University of Vienna's Institute of Hygiene. In 1900 he started mixing samples of his own blood with samples taken from his colleagues, and in the following year he was able to announce his finding. Each person has blood that falls into one of three categories, or 'blood groups', which he called A, B and C. (The C blood group was later re-named the O group, and an AB group was also discovered the following year.) The difference between the groups lay in molecules on the surface of the red blood cells known as 'agglutinogens', and molecules in the serum (the liquid part of blood) called 'agglutinins'. There are two types of agglutinogen, A and B, which are associated with two types of agglutinin, anti-B and anti-A. When an A agglutinogen encounters an anti-A agglutinin (or when a B encounters and anti-B) they bind to each other. This causes the red cells to clump together in a mass. The people with Blood Group O have no agglutinogens, but they have both agglutinins. The whole system is controlled by just one gene.

Landsteiner worked this system out by seeing whose blood was compatible with whose. Its upshot is that people can receive blood from anyone with their own blood group, and also anyone can receive blood with Blood Group O. In Europe, rather over forty percent of people have Blood Group A, slightly more have Blood Group O, while Blood Groups B and AB are rare. However, these ratios are different in other parts of the world. The native Indians of Peru almost all have the same blood group, which is why the Incas were able to practise blood transfusions successfully five hundred years ago.

The discovery of blood groups immediately made blood transfusion practical, and in 1914 a means was devised to store donated blood for long periods without it clotting. This came just in time to be used in the First World War, when blood transfusion saved countless lives, particularly among the Allies. (Strangely, the German medical establishment was much slower to accept it.) Landsteiner, meanwhile, continued research on the subject for the rest of his life, making several important discoveries, including 'rhesus factors', which are important in some life-threatening conditions of newborn babies.

K. Landsteiner, *Wien Klinisches Wochenschrift*, 14, 1132–1134 (1901)

VITAMINS

Necessary components of a healthy diet

By the beginning of the twentieth century, it was already known that any animal's diet needed three major components: proteins, fats and carbohydrates. Further, proteins were already known to be compounded from simpler substances called amino acids. It was widely assumed that the science of dietetics had already achieved its main aims, and only minor details remained to be filled in. However, the whole science of biochemistry was only just beginning, and one of its founding fathers soon made a discovery that served to underline just how much there was still to learn.

The career of Frederick Gowland Hopkins started relatively late by scientists' standards—he did not even enrol as a student until he was twenty-seven years old, in 1888. In 1906 he was researching on the chemistry of proteins, and was feeding rats on diets consisting entirely of purified amino acids, fats and starch (carbohydrate). Rats had been fed on synthetic diets before, but these had always contained impurities. Hopkins managed to synthesise diets free from impurities, but he feared that the result would be unappetising. So to start with he added small quantities of meat or yeast extracts, purely as flavouring. He soon found this was unnecessary—his rats had a hearty appetite for the synthetic diet without any additives being needed. But he also noticed something else. The rats fed on the pure synthetic diet lost weight, while those fed on a diet with even minute amounts of yeast extract thrived. It seemed that the received wisdom was mistaken: proteins, fats and carbohydrates are not sufficient in themselves to form a healthy diet. There must also be other substances as yet unknown, which only needed to be present in minute amounts. So in 1912 Hopkins did a formal experiment: he kept two populations of rats, one fed with a synthetic diet designed to mimic milk, the other fed with the synthetic diet plus a little real milk. The result was clear. There was something in the milk that was necessary for the rats to grow and remain healthy. Hopkins devoted much of the next ten years to research on this mysterious component, and was able to identify several substances he called 'vitamins'.

Hopkins had considerable difficulty in persuading the medical establishment to accept the reality of vitamins. The study of dietetics has always been bedevilled with faddists peddling crackpot theories, and many scientists thought Hopkins was just one more of these. The point at which vitamins became accepted as orthodoxy was probably in 1917, when the manufacturers of margarine asked Hopkins to analyse its nutritional value, compared with butter. He reported that it was lacking in vitamins A and D, and in 1926 margarine with added vitamins was launched on the market.

Hopkins' subsequent career was distinguished. He was the first Professor of Biochemistry at Cambridge University, where his department became a major centre of excellence in both research and teaching. In 1931 he was made President of the Royal Society. He died in 1947, aged eighty-six.

F. G. Hopkins, *Journal of Physiology*, 44, 425 (1912)

INSULIN

The diabetes hormone

Diabetes mellitus has been a well-known condition ever since the seventeenth century. Until the 1920s, it was always debilitating and generally fatal. The discovery of its cause, and how to treat it, was the first time that 'pure' research in physiology yielded a major medical benefit.

By the end of the nineteenth century it was already known that diabetes is an inability to metabolise sugar. It had also been shown, by means of some rather unpleasant experiments on dogs, that it could be caused artificially by removing an animal's pancreas. In 1900 Eugene Opie studied tissues from the pancreas of diabetes sufferers, and could see under the microscope defects in little structures called the 'islets of Langerhans'. He rightly suggested that these 'islets' produced a hormone of some sort, and diabetes was caused by a deficiency of this hormone.

It was from this starting-point that Frederick Banting at the University of Toronto decided in 1921 to see if he could extract the hormone, and use it as a cure. His original experiments were quite crude: he simply removed the pancreas from some dogs, and then endeavoured to keep the dogs alive by injecting them with extracts from another dog's pancreas. The main problem was in producing a suitably refined extract. The pancreas is also a digestive gland, which produces enzymes that pass down a duct into the intestine, and these had to be removed from the extract before it could safely be injected. Banting's original idea was to block off the duct leading to the intestine, and so cause the tissues that generate the digestive enzymes to die off, leaving only the hormone-producing islets. This was later found to be unnecessary, and even counter-productive. Success only came slowly, and it was several months before Banting and his student assistant Charles Best announced that they had succeeded in keeping dogs alive for several weeks without a pancreas. However, the side-effects were severe, and the failure rate was high. The extract they used in their injections was anything but a pure hormone. It was the work of a third member of the team, Professor J. B. Collip, to refine an extract of cow's pancreas to the point where it could safely be used to treat diabetes in humans. The essential ingredient is the hormone insulin.

The team worked extremely hard through the winter of 1921–1922, and the strain showed. They quarrelled violently with each other and with the Head of their laboratory, J. J. R. Macleod. The Nobel Prize was eventually awarded to Banting and Macleod, although Macleod had not personally participated in the experiments. Banting immediately showed his feelings by giving half of his prize money to Best, and Macleod similarly gave half of his prize to Collip. Meanwhile further controversy arose when it was found that a Romanian, Nicolas Paulesco, had been doing similar experiments, and had at one point been rather ahead of Banting and Best. However, he never got as far as refining insulin for treating human patients.

F. Banting and C. Best, *Journal of Laboratory and Clinical Medicine*, 7, 251–266 (1922)

BRAIN WAVES

The electroencephalogram

The human brain is quite easily the most complex single structure that we know of. This makes it the most fascinating thing to study, but also the most difficult. The chief problem lies in how to record its workings in a way that does not impair its function. In addition, researchers are inevitably restricted by very strong ethical considerations. So it is not surprising that up until the 1920s the brain was considered a 'black box' whose internal workings were a complete mystery. Then in 1929 an unknown German professor called Hans Berger demonstrated a way of monitoring the brain's activity: the electroencephalogram, or EEG.

Berger was a shy and rather aloof man, who had become interested in neuropsychology in the 1890s. At that time, research into the brain and the mind were the provinces of two distinct camps: the neuroanatomists, who dissected brains, and the psychoanalysts such as Freud and Jung. Unimpressed by both these disciplines, Berger decided that the best way to study the brain was to find a way to make a physical recording of its activity. He believed that just as the beating of a heart could be recorded by its electrical activity (the 'electrocardiogram'), so the workings of the brain must also produce an electrical signature. His persistence was almost incredible. He continued experimenting on his own for nearly thirty years with barely a hint of success before making the final breakthrough in 1929.

By then, Berger was Director of the Psychiatric Clinic at the University of Jena, where he managed his staff with a punctilious, almost military style. Unknown to most of them, he stayed late in the clinic each evening to do experiments on the patients, attaching electrodes to their scalps in an attempt to record any currents being generated inside their brains. His first success came with patients who had had parts of their skulls removed in order to operate on brain tumours. However, he soon managed to get recordings from subjects with intact skulls (his teenage son was the first of these). He found that when a person is awake but not concentrating on anything, with their eyes shut, the brain produces rhythmic oscillations, now known as the 'alpha rhythm'. Further recordings from patients at the clinic showed how this rhythm was affected by various mental illnesses, especially epilepsy.

Berger's research won international acclaim in 1934, when two Cambridge scientists, Edgar Adrian and H. C. Matthews, using much more sophisticated equipment, confirmed his findings and even pinpointed the precise region of the brain that produced the alpha rhythm. The EEG opened up the whole field of modern brain research, with all the potential it brings for diagnosis and treatment of mental illness. However, in Germany Berger was ignored and even ridiculed. He was forcibly retired from his post in 1938, and his laboratory was closed down. Unable to do research, and horrified by the Nazi regime and the war, he sank into deep depression and committed suicide in 1941, at the age of sixty-eight.

H. Berger, *Archiv fur Psychiatrie und Nervenkrankheiten*, 87, 527–570 (1929)

PENICILLIN

Chance discovery leads to medicine's greatest leap forward

It can fairly be said that until the 1920s there were only two medicines for treating infectious diseases that actually worked: quinine for malaria and arsenic for syphilis. Against other infections, there was little to be done except to go to bed and let nature take its course. Then within a few years of each other, researchers found three potential lines of attack against bacteria: bacteriophages, sulphonamides, and finally antibiotics. Classical antibiotics are substances, produced in nature by moulds and other simple organisms, which kill bacteria by weakening their outer covering so that they burst open. The first to be found was penicillin, which was discovered almost completely by accident by Professor Alexander Fleming in 1929.

Fleming had started research on bacterial infections during the First World War, when soldiers were dying in huge numbers from infections in their wounds, especially gangrene. At that time, the only treatment was to wash the wounds with antiseptic, usually carbolic acid, which is almost as poisonous to the patient as it is to the bacteria. After the War ended, Fleming continued his research at St. Mary's Hospital in London. In particular he looked for something to attack the bacterium *Staphylococcus aureus,* a ubiquitous germ that merely causes boils in healthy people, but which can be life-threatening to a sick patient recovering from an operation. He grew colonies of these bacteria in the usual way, on gelatin in glass dishes. The colonies appear as little round yellow blobs, each about the size of one of the letters on this page.

One day in September 1928, Fleming noticed that one of his dishes contained an unwanted intruder: a patch of mould was growing near the edge of the gelatin. Accidents like this happen even in the best-run laboratories. But he also noticed something else. Around the mould, the *Staphylococcus* colonies were dying off (*see* Plate 13). It only took a few simple experiments to confirm that the mould was producing a substance that could kill bacteria, and moreover was harmless to other living things. The mould was identified as a rather uncommon species called *Penicillium notatum*.

Fleming spent several years trying to use extracts from this mould to treat bacterial infections, but found it very difficult. The bacteria-killing substance broke down very quickly, and lost its efficacy. It was not until the mid 1930s that a special research team was assembled at Oxford University under the leadership of Professor Howard Florey, to develop methods to refine and concentrate an extract of *Penicillium* in a form that remained potent for long periods. They achieved success just in time to make their 'penicillin' available for use in the Second World War. As a result, the Western Allies did not suffer the huge losses from gangrene that they had feared.

Today, penicillin and the other antibiotics are causing some concern. They have become so widely used that bacteria are evolving that have acquired a resistance to them. However, despite these worries, antibiotics remain one of medicine's biggest success stories, and continue to save huge numbers of lives.

A. Fleming, *British Journal of Experimental Pathology*, 10, 226 (1929)

SULPHONAMIDES

The first effective anti-bacterial drugs

It is popularly believed that penicillin, discovered in 1928, was the first antibacterial drug. However, as is so often the case, popular wisdom is only partly right. Penicillin was not successfully made into a drug until the late 1930s, by which time another class of drugs, called the sulphonamides, were already in use. In many ways the discovery of the sulphonamides was the more exciting breakthrough.

The story really started in 1909 when the German chemist Paul Ehrlich invented Salvarsan, a drug to cure syphilis, by modifying a dye that stained the syphilis bacteria so that its molecules contained poisonous arsenic. Salvarsan was not effective against other diseases, but nonetheless it seemed quite possible that other dyes might turn out to have wider anti-bacterial properties. So in the late 1920s the giant German chemical manufacturer I.G.Farbenindustrie, which produced a wide range of synthetic dyes, started a research programme to see if any of its products could be modified to give them medical potential. They recruited Gerhard Domagk, a young lecturer in pathology at the University of Munster, and in 1932 he finally found what they were looking for. He showed that a red dye called 'prontosil rubrum' could protect mice and rabbits against staphylococci and haemolytic streptococci, two kinds of bacteria that can cause fatal blood poisoning. It is not clear why Domagk tried it out on animals at all, since he had already found that it was quite ineffective against the bacteria when they were growing in a culture bottle. Probably he was just being particularly thorough. However, there were rumours that his employers had long suspected the dye's potential, and had instructed him to concentrate his research on it.

Whatever the reason, he soon had good cause to be glad of his persistence, because shortly afterwards his young daughter fell ill with acute streptococcal blood poisoning. As she lay dying, in desperation he gave her a massive dose of prontosil, which cured her immediately. On the strength of this success, I.G.Farbenindustrie quickly patented the drug, and meanwhile Domagk published a full account of his discovery in 1935. Its fame rapidly spread, especially after it was used to save the life of President F. D. Roosevelt's son in 1936.

Domagk was awarded the Nobel Prize in 1939. However, Hitler had banned Germans from accepting Nobel Prizes (a Nobel Peace Prize had recently been awarded to one of his fiercest critics), and so he was unable to receive it until 1947, by which time the prize money was no longer available. By then, prontosil's mode of action had been worked out. It turned out that only half of the molecule was involved, which was why it worked in animals, but not on bacteria growing in a bottle: an animal's body digests the dye, breaking the molecules apart and liberating the active component. This component, called sulphanilamide, became the first of the sulphonamide drugs. They have been used to treat not only blood poisoning, but also pneumonia, gangrene, and several other conditions, and have saved countless lives.

G. Domagk, *Deutsche Medizinische Wochenschrift*, 61, 250–253 (1935)

REJECTION OF TRANSPLANTS

Beginning of modern immunology

Of all the medical advances of the late twentieth century, few were as immediately spectacular as transplant surgery. From the 1970s onwards, the replacement of hearts, lungs, livers and kidneys became almost routine. Previously, such operations had been unthinkable. However, the research that made transplantation possible had its origins much earlier, during the Second World War.

Peter Medawar was a young zoologist at Oxford when he first became interested in immunology. The event that changed his career was an unsuccessful attempt to save the life of an Air Force pilot who had suffered extreme burns. Medawar was left asking himself the question: Since it is possible to give a person a blood transfusion, why is it not possible to give a skin transplant as well? Skin transplanted from another person simply dies and falls off. Medawar decided to make this his full-time area of research, and took a post at the Glasgow Royal Infirmiary, where he worked with the surgeon T. Gibson. In 1942 they performed an experiment which set Medawar on the track that eventually led to our modern understanding about how the body recognises the difference between its own tissues and those of other individuals.

Gibson was treating a young woman who had received a very large burn covering much of her left side. However, unlike the unfortunate airman, enough of her body was undamaged for it to be possible to take several small discs of her own skin and graft them onto about half of the burnt area. These all grew well. The rest of the burn was then given two sets of skin grafts donated by her brother, the second set being grafted on fifteen days after the first. As predicted, these were unsuccessful. The first set grew for about fifteen days before dying; but then the second set failed to grow at all. Medawar immediately recognised the pattern. It was the same as when one encounters a disease organism like the chickenpox virus, for instance. On the first occasion, one has a fever as one's body spends several days 'learning' to recognise the virus and produce the appropriate antibody. On subsequent encounters, one is immune, because one's body recognises the virus and immediately produces the antibody to bind to it. So Medawar reasoned that the rejection of skin grafts also must be an immune response, caused by one's body attacking the foreign skin with antibodies.

Medawar followed up this discovery with some twenty years of research, which earned him a Nobel Prize and ultimately led to the development of the immunosuppressant drugs used in today's transplant surgery. His research career sadly ended in 1969, when he had the first of several strokes that left him increasingly paralysed. However, he then found fame as an author of books about science and the way scientists think. Written in an easy-going style, but always showing the utmost intellectual rigour, they were widely read. Medawar was a warm-hearted man, a devotee of opera and village-green cricket, and much liked by his colleagues. He died in 1987.

T. Gibson and P. B. Medawar, *Journal of Anatomy*, 77, 299–310 (1943)

SICKLE CELL ANAEMIA

The killer disease whose gene protects against malaria

In any list of the greatest scientists of the twentieth century, Linus Pauling must rank highly. In 1932 he united the sciences of physics and chemistry, by showing how the chemical bonding theory of Gilbert Lewis was a consequence of the behaviour of electrons as described by Niels Bohr. In the early 1950s he made major breakthroughs in studying the structure of protein molecules. In 1953 he was close to finding the structure of the DNA molecule when Francis Crick and James Watson beat him to it. And in the late 1940s he led a project that found the cause of one of the world's major health problems, sickle-cell anaemia.

This disease is a debilitating and often fatal congenital condition, in which the sufferer has deformed red blood cells. Many of the cells are crescent-shaped, instead of the normal disc shape. Red blood cells contain the protein haemoglobin, whose function is to carry oxygen from the lungs to other parts of the body. In people with sickle-cell anaemia, the blood does not transport oxygen as efficiently as normal, and the deformed cells can cause obstructions in blood vessels. The spleen is progressively destroyed, resulting in reduced resistance to infections. The disease is almost entirely confined to people of African ancestry, among whom it is common.

Pauling became interested in sickle-cell anaemia in 1945, while serving on a US government committee reporting on what scientific research should receive priority after the Second World War ended. One evening during dinner, sickle-cell anaemia was discussed, and one committee member mentioned that the deformed cells only appeared in blood flowing into the lungs, never in blood leaving them. Pauling realised immediately that the deformity must be due to a defect in the haemoglobin, since this protein is the one thing that undergoes a change as blood passes through the lungs and absorbs oxygen. He at once set up a research project, and was soon able to confirm this insight. Most interestingly, he discovered that the parents of sickle-cell sufferers also had slightly unusual haemoglobin. It was clear that sickle-cell anaemia is caused by a 'recessive' gene: in order to develop the condition, one must inherit it from both parents. If one inherits the gene from only one parent, it only causes very mild symptoms. Pauling learnt that in Africa people with sickle-cell anaemia usually die young, but they seem immune to malaria. He concluded that the gene is common in Africa (where malaria is a major killer) because if both parents in a family carry one sickle-cell gene each, then one-quarter of their children will inherit two sickle-cell genes and die young; another quarter of them will inherit no sickle-cell genes and die of malaria; but half of their children will inherit one sickle-cell gene each and will be resistant to malaria while having only very mild sickle-cell disease.

There is no complete cure for sickle-cell disease, but with the knowledge gained from Pauling's discovery, doctors can advise patients on diet and life-style so as to minimise its effects.

L. Pauling et al, *Science*, 110, 543–548 (1949)

SMOKING CAN CAUSE LUNG CANCER

Clear evidence from hospital statistics

During the first half of the twentieth century, the incidence of lung cancer in the Western world increased enormously. Various causes were suggested, but one attracted particular attention: the huge increase in the amount of tobacco smoked, especially in the form of cigarettes. From the late 1930s onwards, reports started to appear showing that the incidence of smoking among patients admitted to hospital with lung cancer was higher than among patients with other diseases. These early reports only described small-scale studies, but in 1950 two much larger studies were published, which are now seen as the turning-point when the link was firmly established.

The first of these studies to appear in print was conducted by the Americans E. L. Wynder and E. A. Graham. They found that among 605 men admitted to hospital with lung cancer, less than two percent were non-smokers, and over fifty percent smoked more than twenty cigarettes a day. By comparison, among 882 patients with other diseases, nearly fifteen percent were non-smokers, and less than twenty percent smoked more than twenty cigarettes a day. Such a strong association between smoking and lung cancer could not be ignored.

The second study, which was actually begun earlier, was by Professor Bradford Hill and Dr. Richard Doll, working for the Medical Research Council in Britain. At twenty hospitals in

London, the almoners interviewed 649 men and 60 women with lung cancer, and similar numbers of patients with other diseases. They asked these patients various questions about their smoking habits. The results were again quite clear: lung-cancer patients were more likely to be smokers than other patients were, and the correlation was strongest among patients who smoked more than twenty cigarettes per day. This study was more detailed and exhaustive than Wynder and Graham's, and Richard Doll in particular is often credited with being the discoverer of the link between smoking and lung cancer. He followed this original project with several more on the subject, most notably a study of smoking and the incidence of lung cancer among doctors. This was a clever move, because it made the medical profession sit up and take notice in a way that it might not have done otherwise.

Hill and Doll were careful not to claim that 'smoking causes lung cancer'. Plenty of smokers do not get cancer, and many cancer patients are not smokers. Moreover, in 1950 it was not fully appreciated that the tar in tobacco smoke is a carcinogen. This caution has been seized on eagerly by the tobacco industry and the cigarette-smoking public, who have wished to dismiss their claims. During the rest of the twentieth century, the major tobacco companies sought to retain the public's confidence by conducting their own research projects. However, they were seen to have lost the argument when it was revealed that they had suppressed publication of results showing that tobacco smoke does indeed strongly promote cancer. In America, this revelation has left the manufacturers vulnerable to legal action which may yet deal the industry a crippling blow.

R. Doll and A. B. Hill, *British Medical Journal*, 1950/2, 739–748 (1950)
E. Wynder and E. Graham, *Journal of the American Medical Association*, 143, 329 (1950)

THE POLIO VACCINE

The biggest-ever medical experiment

Poliomyelitis, commonly called polio or infantile paralysis, was the last great plague of the Western world until the arrival of AIDS. In North America it reached epidemic proportions: in the 1940s and 1950s as many as one in every five thousand Americans was crippled or killed by it. A race developed to produce a working vaccine.

The winner was Jonas Salk. The son of Polish-Jewish immigrants, he had risen to become Head of the Virus Research Laboratory at Pittsburgh University. He made his breakthrough by combining two recent discoveries made by other researchers. In 1949 John Enders had found a way of growing the polio virus in a culture of tissues taken from the kidneys or testicles of monkeys—work which earned him a Nobel prize. Meanwhile, Salk's colleagues at Pittsburgh had found that the virus could be killed by treating it with formaldehyde. The dead virus, when injected into an animal such as a mouse or a monkey, would cause the animal to produce antibodies just as if it were alive, making the animal immune to subsequent infections. Salk used Enders' technique to grow large quantities of polio viruses (there are in fact three different strains), and killed them with formaldehyde to make a vaccine that could safely be injected.

Salk first tried the vaccine on himself and his family. Satisfied that it seemed safe, in 1952 he vaccinated 39 school children, and then compared the amount of polio antibodies in their blood with the amounts found in people who had recently been paralysed by polio, and also in people who had never had polio. Sure enough, vaccination raised the levels of antibodies in the blood to levels comparable with polio victims. On the strength of this result, the National Foundation for Infantile Paralysis then commissioned the largest-ever medical trial in history. Six hundred and fifty thousand school children from all across the USA were injected either with Salk's vaccine or a placebo, and were then monitored to see if they contracted polio. A further 1.18 million children were monitored as controls. The result clearly showed that the vaccine gave protection, and in 1955 a nation-wide vaccination programme was started.

Salk was lionised by the American public and the media, but the medical and scientific communities were less enthusiastic. He was seen as a self-publicist who used other people's discoveries without giving them the credit. They were much more impressed by his rival Albert Sabin, who was researching on the use of live viruses that had been 'attenuated', that is crippled to make them harmless. This work led in 1961 to a vaccine that can be soaked into a lump of sugar and swallowed. It has now largely replaced Salk's original version.

Salk continued to work on virus research until his death in 1995. He also set up the prestigious Salk Institute in California, an independent laboratory where many leading biologists have been able to do pure research, without the responsibility of university teaching or the need to produce results for employers in industry.

J. Salk, *Journal of the American Medical Association*, 151, 1081–1098 (1953)

INTERFERONS

Hope for treating viruses and tumours

Most infectious diseases are caused either by bacteria or by viruses. The arrival of antibiotics and sulphonamide drugs in the mid-twentieth century marked major victories over bacterial diseases, but as yet most virus infections remain incurable. However, in 1957 a team led by Dr. Alick Isaacs at the National Institute for Medical Research at Mill Hill, on the outskirts of London, made a discovery that may yet prove to have been a major turning-point: a chemical substance that our bodies naturally generate when fighting a virus infection.

Isaacs and his team were studying the phenomenon of 'interference' in virus infections. When we contract a virus disease, it temporarily protects us from further infection. It is unusual to catch two virus diseases at once. Apparently when virus particles invade a living tissue, something happens in that tissue that gives it protection from further invasion. Isaacs was studying this effect, using pieces of membrane from inside a developing chicken's egg, a tissue that is easy to grow in a test-tube. He started by exposing the membrane samples to influenza virus that had been 'attenuated', that is rendered innocuous, by heating it to 50 °C for an hour. Then he washed the attenuated virus away, and exposed the membrane samples to live, infectious virus. He found that if the membrane samples were exposed to attenuated virus for just five minutes, a resistance to live virus would then build up over the next hour or so, provided the samples were kept warm. He reasoned that the attenuated virus must be stimulating the membrane to start generating a substance that interfered with other viruses. He called it 'interferon'.

Research on interferon progressed quietly for the next twenty years or so. It was found that there were three different kinds, called alpha, beta, and gamma, and they are all proteins. In humans, they are mainly produced by certain white blood cells. Then, around the late 1970s and early 1980s, the trickle of research publications suddenly turned into a flood. The latest advances in biotechnology meant that synthetic interferons could be manufactured in large quantities, and suddenly there was the prospect of using them as drugs for treating virus infections.

Today, it seems that interferons have not lived up to their early expectations as wonder drugs. They work by triggering a variety of defensive mechanisms in living tissues, which makes them almost as poisonous to the tissues that produce them as to the virus itself. Indeed, they are at the root of many of the symptoms of fever. Nonetheless some persistent viral infections, notably hepatitis B and C, do yield to interferon treatment, especially in combination with other drugs. Recently it has also been found that interferons may be useful for treating a variety of diseases that are not caused by viruses. Several kinds of cancer, including some leukaemias, multiple sclerosis, and possibly diabetes and rheumatoid arthritis, all seem to respond to interferon treatment, at least temporarily. So, although no panacea, synthetic interferons may yet prove to be a major weapon in the medical armoury.

A. Isaacs and J. Lindemann, *Proceeding of the Royal Society of London B*, 147, 258–267; A. Isaacs et al, *Proceedings of the Royal Society of London B*, 147, 268–273 (1957)

PRIONS

The killer molecules of BSE

In the last two decades of the twentieth century, the British public became uneasily aware that a new and horrible disease was striking down our cattle. 'Mad Cow Disease', or Bovine Spongiform Encephalopathy (BSE), to give it its full name, caused cows to lose all control over their bodies, to become demented, and finally to die. The Government assured everyone that the disease could not be passed on to humans, only to be proved wrong when the first human fatalities from BSE started to appear in 1995. It was widely (and quite possibly wrongly) reported that BSE had originally entered cattle through artificial cattle-feed containing the rendered remains of sheep carrying a similar disease called 'scrapie'. The disease was then spread via cattle-feed made from the remains of infected cows. Moves to keep infected meat out of animal feeds came too late to prevent a major epidemic that has still not died out, and is now spreading to other countries.

The Government was widely lambasted for its complacency, but the scientists who advised it had drawn perfectly natural conclusions from the evidence they had at the time. Scrapie had been endemic in sheep for centuries, during which time many an infected sheep must have ended up as mutton. And yet, scrapie had never appeared in people. So if one cannot catch scrapie from eating infected sheep meat, it seemed safe to assume that one could not catch it from eating infected beef either. However, it now appears that BSE is not actually scrapie, but rather a separate disease, which is capable of spreading from one species to another.

The problem with scrapie and BSE is that unlike other infectious diseases they are not caused by bacteria or viruses, but rather by something very strange called a 'prion'. The extraordinary nature of prions was shown by Stanley Prusiner and his colleagues as long ago as 1983, but their findings were so weird that until very recently there have still been scientists who have refused to believe them. A prion is simply a transformed version of a protein found in all animals' nervous systems. If even a tiny amount of this transformed protein gets into an animal's nervous system, it causes all the normal protein to change into the transformed version by some kind of domino effect. Nerve cells containing the transformed protein become deformed, so that the infected animal's brain progressively stops working. The disease is invariably fatal.

At the start of the twenty-first century, it was still unclear how serious the epidemic of BSE in humans was going to be. It appeared that not everyone was susceptible to it, and it seemed likely that eating infected meat did not always automatically result in infection. But it also became clear that the precautions taken to keep infected tissues out of the butchers' shops had not always been followed carefully. The disease has an incubation period of several years, and it is unlikely that we will know the worst for some time yet.

S. Prusiner et al, *Cell*, 35, 57–62 (1983)

HIV

The AIDS virus

The appearance in the early 1980s of AIDS as a major killer was arguably the biggest calamity to befall the human race in the twentieth century. In Africa alone, the death toll is already running into tens of millions. At first, many people doubted that it was caused by a disease organism at all. (A few people still persist in these doubts, but the overwhelming bulk of scientific evidence is against them.) However, to medical researchers a virus always seemed the most likely cause, and several laboratories were soon at work trying to identify it.

It was almost immediately found that AIDS patients were losing certain white blood cells called T Lymphocytes, which are a vital part of the immune system. As a result, their immune systems collapsed, leaving them susceptible to every disease organism they encountered. This discovery attracted the interest of Robert Gallo at the US National Cancer Institute laboratory on the outskirts of Washington, DC. He had recently discovered two viruses that attacked T cells, although these viruses, rather than killing the T cells, caused them to start multiplying uncontrollably, resulting in leukaemia. Meanwhile, Luc Montagnier at the Institut Pasteur in Paris also started a virus hunt. At first, the two laboratories collaborated, and in 1983 they simultaneously published articles in *Science* magazine claiming to have found evidence that AIDS was caused by a leukaemia-type virus. The article from Montagnier's team (its principal author was Françoise Barre-Sinoussi) even contained a photograph taken with an electron microscope, showing the virus particles. It was a 'retrovirus'—one that carries its genes in a molecule of RNA rather than the usual DNA. When it infects a T cell, it makes copies of its genes in DNA, which then get incorporated into the T cell's genome.

After the initial discovery, the story turns ugly. The French and American teams ended their close collaboration, and a race developed as each laboratory tried to prove that its own virus strain was indeed the cause of AIDS, to prepare viable cultures, and to sequence its genome (i.e. its entire set of genes). The winner would reap not only scientific kudos, but also patent rights to the technology for diagnosing infection with the virus. At one point Gallo was accused of fraudulently presenting Montagnier's virus samples as his own. An enquiry by the US Office of Scientific Integrity was unable to prove deliberate fraud, but strongly criticised his laboratory for poor techniques and lack of adequate record-keeping. Eventually the US and French governments came to an agreement to share the disputed patent rights.

Gallo and Montagnier are today acknowledged as the joint co-discoverers of the virus now known as HIV. Gallo's early belief that it was similar to the leukaemia viruses turned out to be inaccurate. Rather, it belongs to the 'lentivirus' family of retroviruses, which were previously known to cause slowly-developing diseases in farm animals. HIV probably first infected humans in western Africa in the 1930s. Similar viruses are found in apes, and it may have entered the human race through a fight between a hunter and a chimpanzee.

F Barre-Sinoussi et al, *Science*, 220, 868–871 (1983)

HELICOBACTER PYLORI

The stomach ulcer bug

Gastric ulcers are a well-known complaint. In a healthy person, the stomach lining secretes juices containing a high concentration of hydrochloric acid, which helps to sterilise our food before it passes on into our intestines to be digested. The stomach lining itself is protected from the acid by a layer of mucus. However, this mucus sometimes fails to do its job adequately, with the result that the stomach and the duodenum (the top end of the intestine) suffer acid burns. They are very painful and debilitating, and sufferers often cannot eat without vomiting. It has long been noted that ulcers are commonest in people who lead stressful lives, and so the traditional treatment has been to lie quietly in bed and subsist on a diet of milk until the pain stops. This is effective but slow, and it does not prevent the ulcers from recurring in future. A much quicker treatment became available in the late 1970s, with the arrival of drugs that reduce the amount of acid secreted into the stomach. These drugs could stop the symptoms in a day or two, but again there was no guarantee that the cure was permanent. (The drugs have proved immensely profitable to their manufacturers in consequence.)

However, a few doctors always had suspicions that excessive acid was not the real cause of ulcers. Among these was Barry Marshall of Freemantle Hospital in Australia, who was convinced that a bacterial infection was involved. His belief met with general scepticism until 1985, when he proved it by performing a simple experiment on himself. He isolated some bacteria from the stomachs of patients with gastritis, an inflammation of the stomach lining that often precedes fully-blown ulcers. He grew them in a culture, and when the culture was growing well, he swallowed it. Sure enough, after a week he developed gastritis, which lasted for about a fortnight. Samples taken from his stomach lining showed that the bacteria had embedded themselves under the mucus layer, which had become much thinner than normal. Marshall then followed up this experiment with surveys of patients with stomach ulcers, and found that the vast majority of them had the bacterium. Among people without ulcers, however, it was much rarer.

The bacterium, which has since been named *Helicobacter pylori*, is extremely widespread. In the affluent western nations it is in decline, but in the developing countries it is likely that most people carry it. Most people do not get ulcers, of course, so *H. pylori* is clearly not 'the cause' of ulcers by itself. However, it has been found that ulcers can be cured by attacking the *H. pylori* with antibiotics, and what is more, this cure is permanent. *H. pylori*, it seems, is harmless enough under normal circumstances. But sometimes, especially when one suffers long-term stress, it seems that one's body can no longer cope with it, and it is then that ulcers develop. A less stressful lifestyle would seem to be the best cure, but failing that, removing the bacterium will stop the stress from causing ulcers.

B. J. Marshall et al, *Medical Journal of Australia*, 142, 436–439; 439–444 (1985)

EMBRYONIC STEM CELLS GROWN ARTIFICIALLY

Prospects for reversing degenerative diseases—or for genetically modified humans

It has long been known that an early embryo consists in large part of a mass of cells, all seemingly the same, and apparently not specialised to perform any particular function. These are the 'embryonic stem cells'. Their importance to medical research lies in the fact that under the right conditions, any kind of tissue can be grown from them. Older foetuses also contain stem cells, but these become more and more committed to growing into specific tissues. For example, nervous tissue stem cells taken from aborted foetuses have recently been used to treat Parkinson's disease, by placing some of them in the patient's brain. (So far this treatment has yielded rather mixed results.) Embryonic stem cells, by contrast, are not committed to grow into any particular tissue.

However, in order to use embryonic stem cells for medical purposes, they must first be grown artificially in large quantities. This was finally achieved in 1997, when John Gearhart of Johns Hopkins University, at Baltimore, Md, managed to culture stem cells taken from foetuses that had been aborted around seven weeks after conception, and kept the cultures alive and growing for several weeks. His announcement caused major excitement, but strong misgivings were also expressed by a number of people, including several scientists. It was immediately realised that although Gearhart himself was only interested in using his discovery to treat diseases, it could possibly also be used for producing genetically modified humans, or even for cloning people.

Growing embryonic stem cells in culture is not new. It had first been achieved with stem cells from mouse embryos in 1981 by Martin Evans and M. H. Kaufman at Cambridge University, and also independently by Gail Martin at the University of California, San Francisco. However, their techniques could not be used on humans, because of laws prohibiting experiments on human embryos. This was why Gearhart used tissues from aborted foetuses. In America there is no law against using material obtained from these, so long as they were not deliberately aborted for that purpose. Gearhart identified a small part of the foetus called the 'gonadal ridge', which appeared still to contain embryonic stem cells when the foetus was up to eight weeks old.

Gearhart's achievement caused anxiety because in mice, embryonic stem cells are regularly used to produce genetically modified individuals. It is fairly easy to modify the genome of a stem cell, and if genetically modified stem cells are injected into a normal embryo, it may grow into an adult that bears genetically modified offspring. Few people liked the prospect of this technique being used on humans. However, embryonic stem cells can be genetically modified in such a way that they can be transplanted into an adult person without being rejected. Stem cells modified in this way could have enormous potential for treating all kinds of degenerative diseases, where an organ or tissue has stopped working and needs to be replaced. The general consensus among medical researchers is that this potential benefit should not be foregone, and that human stem cell research should continue.

J. D. Gearhart, *13th International Congress of Developmental Biology* (1997)

HISTORY OF LIFE

Of all the branches of science, few have had such popular appeal as palaeontology, the study of the history of life on Earth. Of course, fossils have been known about since time immemorial. Until the late eighteenth century, they were generally assumed to be the bones of animals that had been destroyed in Noah's Flood. However, by the early 1800s geologists were already becoming convinced that the Earth was far older than had previously been suspected, and that the rocks with their fossils had been laid down very gradually over an immense period of time. 'Catastrophism', the belief that the whole Earth had been subjected to an immense flood (and probably several other, earlier cataclysms) came to be seen as unscientific, and fell out of fashion. This revolution in the geologists' thinking was followed by a whole spate of fossil discoveries in the nineteenth century, first in England and Europe, and later in North America. These were immediately popular with the general public. Where previously people had dreamt of imaginary monsters such as dragons and unicorns, now they could be excited at the thought of real dinosaurs. Palaeontology was fun.

In the twentieth century, study of the dinosaurs progressed rapidly. Unquestionably the most important discovery has been the final answer to the old question: Why did they become extinct? It is now almost certain that they were all killed when an asteroid hit the Earth, sending up a cloud of dust that blotted out the Sun for years on end. In fact the extinction of the dinosaurs was not the only, nor even the biggest, mass extinction in the Earth's history, and it is possible that at least some of the other extinction events were also caused by asteroids. This discovery has made many scientists uncomfortable. For over a century they had been taught to eschew 'catastrophism'. Now it seems that global catastrophes really have occurred several times in the past, and have greatly changed the course of life's history. Not surprisingly, several palaeontologists have been highly sceptical about them, and have tried to argue that the mass extinctions were really not all that sudden or mysterious. However, these counter-arguments are now wearing thin, as increasing evidence comes in.

Dinosaurs are undoubtedly fun to contemplate, but for most non-scientists they are really just entertainment. By contrast, the questions likely to affect us most deeply are: How did life begin? And where did the human race come from? In earlier times (and throughout much of the world even today), these questions have been the province of religion, and when science turns its attention to them, it inevitably stirs up peoples' emotions. The study of how life began has received relatively little publicity, probably because the scientists involved realise just how little they have to tell. In the early years of the century, the Russian Aleksander Oparin speculated about how non-living chemical substances could have come together to make the first living things, but serious experimentation did not start for another fifty years. Today, a century after Oparin, we are still speculating. All that can really be said is that we now know some non-living processes that will generate the basic molecules from which living things are built. As to what happened next, various ideas are being floated, but we are still largely unable to test them by experiment.

The search for humanity's ancestors, by contrast, has made enormous progress during the last hundred years, and the rate at which discoveries are being made is still accelerating. However, the emotive nature of the subject has repeatedly led to highly regrettable behaviour among the scientists involved. In the first half of the century, palaeontologists showed their racist attitudes by assuming that our ancestors must have lived in Europe and Asia, despite the fact that most other great apes live in Africa. When the weight of evidence finally forced them to acknowledge our African antecedents, they immediately sought to denigrate them, portraying them, on scant evidence, as murderous cannibals. Worse still was the infamous Piltdown Skull, 'discovered' by the palaeontologist Charles Dawson in 1912 in a gravel-pit in southern England, and presented to the scientific community as the 'missing link' between apes and humans. It was not until 1953 that it was finally shown to be a hoax, created by joining the cranium of a human being with the jawbone of an orang-utan. We will probably never know whether Dawson was the actual perpetrator, or whether he was duped by a colleague, but whoever the true culprit was, he apparently wanted to convince people that the human race originated in Western Europe. Alternatively, he may have intended it as a malicious practical joke, designed to make fools of his professional rivals. If so, it would not be the only instance of professional jealousy leading palaeontologists to behave badly. When faced with the bones of our ancestors, even trained scientists are apt to become over-emotional, and the whole field has been beset with the most immoderate squabbles and feuds.

The need for creation myths seems pervasive in human society. In the secular Western world, with the Book of Genesis no longer taken literally, people have often turned to the fossil record to meet this need. In popular books the history of life is repeatedly told as one of straightforward progress, starting with the simplest organisms and leading inexorably upwards towards ourselves. In fact, the evidence suggests a much more haphazard and tangled tale. Our immediate ancestors have also repeatedly been mythologised, even by supposedly serious authors, who have presented them either as savage brutes or as child-like innocents, according to personal preference. The truth is, that science is not religion, and anyone who wishes to turn the fossil record into a moral story will have to make much of it up as they go along.

This section has been arranged differently from the others. In other sections, the discoveries are listed in chronological order. In this section, the discoveries are listed in order of the period in life's history that they relate to, starting with discoveries about very early life, and ending with the most recent fossils.

FIRST CLUES ABOUT HOW LIFE BEGAN

Cooking up the primordial soup

The question of how life arose on Earth is one of the most intractable problems that scientists have set themselves. All known living things need a minimum of two kinds of molecules—nucleic acids (DNA or RNA) to carry genetic information; and proteins to provide the machinery for making copies of the nucleic acids to be passed on to the next generation. But in today's world, the two are never found on their own. Nucleic acids are needed to manufacture proteins, and proteins are needed to make nucleic acids. They are also the most complex of all known molecules, so we are faced with the questions: How did they arise in the first place; and how did they come together? There has been a mass of theoretical work, but experimental evidence is coming very slowly.

The earliest, and most famous, experiment was performed by Stanley Miller at the University of Chicago in 1953. Together with the astronomer Harold Urey, he decided to see if the chemicals in the early Earth's atmosphere could have provided the raw materials from which the first living things were made. In the earliest period of the Earth's history its atmosphere was very different from today. There was certainly no free oxygen, because this is a highly reactive gas produced by green plants. Urey believed that the atmosphere probably consisted of the gases methane (a compound of carbon and hydrogen, the main component of the 'natural gas' we burn in gas heaters), ammonia (a compound of hydrogen and nitrogen), water vapour and hydrogen. Miller mixed these gases together, and placed them in an apparatus where electric sparks were repeatedly passed through the mixture. The idea was to simulate the effect of lightning flashes. After a week of this treatment, Miller analysed the resulting chemicals. A sticky yellow mixture of organic compounds had accumulated at the bottom of the apparatus, which proved to contain significant quantities of amino acids, the molecules from which proteins are built. Miller and Urey suggested that early in the Earth's history, amino acids were formed in thunderstorms, and the rain then washed them into the sea, where they provided the original building-blocks from which the first living things assembled themselves.

It is fair to say that this experiment has often been quoted with excessive hype. Miller had not come anywhere near to showing how life arose. Moreover, it is now believed that Urey had got the composition of the early atmosphere completely wrong. It is now thought to have consisted of carbon dioxide, nitrogen, and water vapour. However, the experiment has since been repeated using this mixture of gases, and the results have been very similar. Amino acids seem easy to make, and even nucleotides, the building-blocks of nucleic acids, can be made by such simple means. The main significance of the Miller-Urey experiment was to show that the question of how life arose could be investigated in the laboratory, and need not forever remain a matter of pure speculation.

S. L. Miller, *Science,* 117, 528–529 (1953)

THE EARLY PRESENCE OF LIFE ON EARTH

Evidence for life nearly four billion years ago

Fossils have always caught the public imagination. Countless books have been published describing the animals and plants of the past, from the trilobites of the Cambrian Period, through the age of the dinosaurs, to the mammoths and sabre-toothed tigers of recent times. However, these popular books tend to give a misleading impression. They seem to suggest that life on Earth began less than six hundred million years ago, and that highly complex animals and plants appeared almost at once. In fact, life has been present on Earth for over six times longer than that, although for most of that time the fossil record is very scant.

It is likely that for most of the Earth's history, life consisted entirely of very simple organisms. Generally they were microscopically small, although they often lived in large colonies. To date, the oldest rocks claimed to contain fossils are in north-western Australia, aged from 3.3 to 3.5 billion years. Marks in these rocks resemble bacteria-like cells, strung together in short chains of perhaps a dozen cells each. However, simple though they are, they must have evolved from even simpler precursors. These precursors would probably not leave recognisable fossils, so we would expect only to find indirect evidence for their existence.

Such evidence was found in 1996, by a team headed by S. J. Mojzsis of the Scripps Institute of Oceanography in

California. They studied rocks on the western coast of Greenland dated at 3.5 to 3.8 billion years, which makes them about the oldest rocks to be found anywhere on the Earth's surface. They contain carbon, and the question Mojzsis asked was: Did this carbon originate in living organisms, or did it enter the rocks by some non-living process? Carbon occurs as a mixture of two isotopes, carbon-12 with six neutrons in each atom and the much less abundant carbon-13 with seven. When living things take up carbon from their surroundings, they absorb carbon-12 more readily than carbon-13. This means that one can tell if the carbon in an ancient rock came from living organisms by the amount of carbon-13 it contains. Mojzsis analysed the ratio of the two isotopes in the Greenland rocks, and found a clear lack of carbon-13. This, he said, showed that the carbon must have passed through living organisms.

Mojzsis' discovery showed that life must have appeared on Earth just as soon as it could. The Earth is 4.55 billion years old. It is believed that around 3.9 billion years ago, it received a massive bombardment of asteroids, which must have melted its entire surface for tens of millions of years. After that, life seems to have appeared very quickly indeed. This opens up an intriguing possibility. Did life really originate here on Earth, or did it arrive from outer space on board an asteroid? This idea has been suggested several times over the last two hundred years, but previously most scientists had dismissed it as being completely crackpot. Since Mojzsis' discovery, however, several researchers have started to take it more seriously.

S. J. Mojzsis et al, *Nature*, 384, 55–59 (1996)

THE ORIGIN OF HIGHER ORGANISMS

How did our ancestors evolve from bacteria? By forming a co-operative

When most people think of living things, they think of creatures big enough to see with the naked eye: humans, trees, and so on. However, these are exceptional, for two reasons. Firstly, they are big—while the huge majority of living things are microscopically small. But more significantly, they all belong to the more complex of the two main kinds of living things—the 'eukaryotes'. The other kind, the 'prokaryotes', are all microscopic, and have a much simpler basic architecture. The most familiar prokaryotes are the bacteria, and they are by far the most numerous of all living things. No biologist doubts that the eukaryotes are descended from prokaryote ancestors. However, the leap in complexity from prokaryotes to eukaryotes is enormous, prompting the question: how did evolution take this giant step?

There are two major differences between the cells of prokaryotes and those of eukaryotes. The first is how their genes are organised. Prokaryotes (which always consist of just one cell) contain only one set of genes, mostly contained on one big loop of DNA. Eukaryote cells, by contrast, usually contain two copies of each gene, and they are carried on a number of bodies called chromosomes. The second difference is that eukaryote cells (which are generally much bigger than prokaryotes) contain various complex pieces of machinery that prokaryotes do not. In particular, they contain little sausage-shaped bodies called mitochondria, which perform the task of extracting energy from food and oxygen. Plants (which are all eukaryotes) also contain little green egg-shaped bodies called chloroplasts, which trap sunlight and use its energy for making sugar out of water and carbon dioxide. Strangely, mitochondria and chloroplasts carry their own genes inside them, and divide and reproduce themselves just like little independent organisms.

It was Lynne Sagan (now Lynne Margulis) at Boston University who first spotted the obvious implication of this fact. Mitochondria and chloroplasts look like bacteria, she said, because that is just what they are. Their ancestors took to living inside other, bigger, bacteria around two billion years ago, at about the time that the Earth's atmosphere started to contain large quantities of oxygen. During the ensuing ecological upheaval, some organisms found a competitive advantage by teaming up, with small bacteria sheltering inside a larger organism and performing its vital chemical reactions. Sagan also noted that the smallest eukaryotes swim by means of a 'flagellum'—a whip-like thread that they thrash to and fro. This looks very like some of the thread-like bacteria, and could also have started by being a separate organism. Sagan suggested that the flagellum's genes went to form part of the eukaryotes' chromosomes, which were evolving at the same time.

Today, Sagan's explanation is universally accepted as being largely correct. However it was slow to spread, and for several years she was considered something of a maverick outsider. Biologists generally think of evolution as being driven largely by competition between organisms, and they found it hard to accept that such a major development resulted from different organisms coming together to co-operate.

L. Sagan. *Journal of Theoretical Biology*, 14, 225–274 (1967)

NEMESIS OF THE DINOSAURS

An asteroid hits the Earth

Dinosaurs have always caught the public's imagination. They first evolved over 170 million years ago, and dominated the animal kingdom for the next 110 million years. Then suddenly, about 65 million years ago, they became extinct, along with about half of all living species of animals and marine plants. (Curiously, most land plants survived.) Many theories were put forward to explain this mass extinction, some of them quite bizarre, before the most likely cause was found.

In 1980 the geologist Walter Alvarez was studying limestones in Italy, along with his father Luis, formerly an eminent physicist, but now officially retired. The rocks they were studying consisted of the fossilised remains of plankton, that had been laid down on an ancient sea-bed around the time of the great extinction. Just at the exact time of the extinction, there was an interruption in the limestone, marked by a layer of clay about one centimetre thick. This clay was found to contain abnormally large quantities of iridium, a rare metal normally associated with asteroids. Alvarez looked at rocks in other locations, in Denmark and New Zealand, and everywhere found the same thing. At the moment that the dinosaurs became extinct, a sudden influx of iridium appeared all over the world.

Alvarez concluded that there was only one feasible explanation for the iridium layer. Sixty-five million years ago, an asteroid with a diameter of ten kilometres or more hit the Earth, causing an explosion equivalent to a hundred million tonnes of TNT, and sending up a cloud of dust that entered the upper atmosphere and spread all over the globe. Sunlight was entirely blotted out for several years. It is also likely that red-hot debris rained down all over the world, killing all land animals too large to take cover. It was remarkable that any living things survived at all.

Not surprisingly, many scientists disputed this lurid scenario. It sounded too much like science fiction, or even mythology. The extinction of the dinosaurs appears sudden, they said, but in reality it was probably spread out over one or two million years. Its most likely cause was major changes in the world's climate, probably triggered by massive volcanic eruptions that were going on at that time in India. However, these critics were largely silenced when geologists announced that they had found the site of the impact. On the coast of Yucatan, Mexico, there is a meteorite crater some 200 kilometres across, now filled with sediment and debris. In 1991 its age was calculated, and it was found to be just 65 million years old. Even the most determined sceptic had to agree that this impact was the most likely source of the iridium layer, and must have caused a worldwide holocaust of living things.

This discovery has prompted scientists to ask: could it happen again? And if we saw an asteroid heading towards Earth, could we prevent it from hitting us? Much of today's research effort into asteroids is directed at these questions.

L. W. Alvarez et al, *Science*, 208, 1095–1108 (1980)

'LUCY'

Skeleton of a pre-human ancestor

Starting in the 1920s, scientists increasingly came to realise that the human race had originated in Africa, rather than Europe or Asia as had previously been assumed. The earliest human species was *Homo habilis*, which lived in eastern Africa some two million years ago. It is presumed to be descended from a genus of apes called *Australopithecus*. There seem to have been at least three species of *Australopithecus*, and possibly a lot more. Some were heavily-built with very strong jaws, while others were more gracile and human-like. However, until the 1970s our knowledge of them was very slight, mostly gleaned from fragments of skulls. Bones from other parts of the body were much harder to interpret, especially as no-one had yet found anything like a complete skeleton.

This changed on November 30, 1974. On that day, Donald Johanson from Cleveland Museum of Natural History was leading a fossil-hunting expedition at Hadar in Ethiopia. Reluctant to face a pile of unfinished paper-work in his tent, he went out with his student assistant Tom Gray to search for fossils in a nearby river-bed. At about noon, in almost unbearable heat, they suddenly started finding bone after bone. Their excitement mounted when they realised that all the bones had belonged to the same individual. They soon had about half of the entire skeleton of a small female *Australopithecus afarensis*, an early species which flourished about four million years ago. They immediately nicknamed the skeleton 'Lucy', after the Beatles' song 'Lucy in the Sky with Diamonds'. It has told us more about our pre-human ancestors than any other single fossil (*see* Plate 10).

The remarkable thing about Lucy is how human she looked. Her head was reminiscent of a chimpanzee's, and she had curved fingers suitable for grasping branches of trees, but the shape of her pelvis and thigh-bones showed that she walked on the ground as upright as a modern human, albeit with rather short legs. So to answer the old question 'When did Man come down from the trees?', we can now say that it was our pre-human ancestors who came down from the trees, and Lucy probably lived at just about the time when they made the transition.

It is tempting also to ask: How close is Lucy to the fabled 'Missing Link' between humans and other apes? However, scientists have long since ceased to talk about a Missing Link, because it is clear that there was more than one. A species called *Homo habilis*, which lived about two million years ago, is probably close to the link between *Australopithecus* and humans. However, we do not yet have anything near to the immediate common ancestor of hominids and our nearest living relatives, the chimpanzees. Genetic analyses suggest that the lines leading to chimpanzees and humans diverged about six million years ago, which is still a blank period in the fossil record. Indeed, we have no fossils of the chimpanzees' ancestors at all—unless, as some have suggested, they are descendants of *Australopithecus* that went back up into the trees again.

D. C. Johanson and M. Taieb, *Nature*, 260, 293–297 (1976)

THE 'TAUNG CHILD'

A pre-human fossil from Africa

The search for fossil hominids—humankind's ancestors—has always excited people. Scientists find themselves being asked questions like 'When did our ancestors come down from the trees?', or 'Have we found the Missing Link?' The quest also stirs the emotions. To this day there are many people, especially in North America, who cannot contemplate the fact that we are descended from apes. Professional researchers, meanwhile, engage in surprisingly heated arguments over how to interpret their fossil finds, and the whole field has been dogged by personality clashes among the relatively few scientists actively involved.

At the beginning of the twentieth century, fossil humans had already been found dating back half a million years. The oldest came from eastern Asia, and it was widely assumed that the ancestors of *Homo sapiens* lived somewhere in Europe or Asia. So when even older hominid fossils started turning up in Africa, most experts refused to believe that they belonged to the human lineage at all. The first of these to be found was the famous 'Taung child', the skull of a young ape with some remarkably human characteristics, unearthed in a limestone quarry in South Africa.

The Taung Child's discoverer, Raymond Dart, was not an expert on fossils, although he had a strong amateur interest in them. He was professor of anatomy at Witwatersrand University, and in 1924 he heard from one of his students of a quarry where the bones of fossil apes were being found. He visited the quarry, and asked its owner to send him any more fossils found there. In due course he was sent a lump of rock from which pieces of a skull could be seen sticking out. He spent over two months carefully freeing the skull from the rock, and at the end he was amazed at what he had found. Dart knew little about fossils, but he knew a great deal about skulls, and this one was extraordinary. It was an ape, but one with an abnormally large and round brain-case. Most astonishingly, the hole where the spinal cord emerged was positioned on the underside of the brain-case rather than at the back, showing that the animal had stood at least partly upright like a human. He named it *Australopithecus africanus* ('African southern ape'). The fossil was originally estimated to be about a million years old, about twice as old as the Asian fossils. It is now believed to be even older—nearer two million years old.

It took some twenty years for the scientific establishment to accept that *Australopithecus* was related to humans. Only when several more similar skulls and other bones were found in southern Africa did opinion turn in its favour. Many of these showed signs of having been butchered, which caused Dart to propound the most lurid theories about our ancestors' savagery and cannibalism. However, it is now believed that most of these fossils are remains of hominids that had been devoured by hyenas and leopards rather than by their own kind.

R. Dart, *Nature*, 115, 195–199 (1925)

BIOLOGY

More than any other branch of science, biology came of age during the twentieth century. In 1900, researchers in biology fell broadly into two categories: physiologists and medical researchers, who could claim to be practising real science with theory, experiment and observation; and 'naturalists', who were still doing little more than cataloguing facts and filling museums with specimens. Darwin and Wallace had proposed a fully scientific theory of evolution some fifty years earlier, but nobody really had any idea of how to test it. By the beginning of the twentieth century almost all serious biologists believed that evolution was a fact, but they did not generally accept that it was being driven by natural selection as Darwin and Wallace had claimed. They knew nothing of genetics, and next to nothing about ecology, the two sciences which between them show how evolution works. Meanwhile, the whole science of biochemistry had yet to begin; the sheer size and complexity of biological molecules still left chemists completely baffled.

A century later, the picture had changed completely. Physiology had progressed, perhaps, fairly steadily. But on either side of it (so to speak), biochemistry, ecology, behavioural and evolutionary sciences had all taken off, together with genetics, whose rise is described in a separate section of this book. In the case of biochemistry, the chief agent of this change was undoubtedly the development of new techniques, especially X-ray diffraction analysis, which made it possible to analyse the structure and composition of large molecules. Physiology also benefited from the development of new methods, notably of growing living tissues in cultures, and also the invention of the electron microscope, which can achieve magnifications far greater than a conventional optical microscope can. However, the rise of the 'whole-organism' disciplines of ecology, behaviour and evolution seem chiefly to have resulted from the efforts of a few pioneers, who determinedly moved their fields away from the simple observing and describing of the natural world, and started on serious analysis and experimentation. Previously, naturalists had merely used words to describe their studies; now they started to use mathematics as well.

However, the achievements of the 'whole-organism' scientists do not feature largely in this book. Von Frisch's analysis of honey-bees' dances and Lack's study of the Galapagos finches are the only two discoveries from these fields to be included. This is because these sciences, even more than others, progress almost entirely in small steps, by the cumulative effect of very many minor discoveries. Giant leaps are few and far between. This is presumably what the geneticist Francis Crick meant, when he reputedly said that 'there is no Nobel-Prize-winning work to be done in ecology'. There have been several major theoretical advances, such as the application of game theory to evolutionary biology by John Maynard Smith in the 1970s, but seeing how such theories apply to real organisms and environments has been a slow and piecemeal business.

The biological sciences inevitably catch the general public's interest, because they apply directly to ourselves. It was equally inevitable that the studies of evolution, ecology and behaviour would cause controversy, and biologists found themselves under continuous attack, mainly from two directions. Firstly, there were the 'creationists', mostly in North America, who could not accept evolution theory at all for religious reasons, and tried to argue that evolutionary biologists were misreading the evidence. It was easy to demolish their arguments from a scientific viewpoint, but this did little to diminish their power-base among people with little or no scientific education. By ensuring that evolution received scant mention in high-school textbooks, they had a serious effect on the way biology was taught. Repeated legal attacks on their position, based on the First Amendment to the American Constitution which bars the teaching of religion in state-sponsored schools, have only partially contained the threat they pose to serious science.

The second attack the biologists faced was over the subject of 'sociobiology'. Two popular books published in the 1970s, *Sociobiology* by E. O. Wilson and *The Selfish Gene* by Richard Dawkins, contrived to give many people the impression that evolutionists took an extreme position in the 'nature-versus-nurture' debate about human behaviour. They were accused of denying the existence of free will, and portraying people as automata with no control over their impulses. These accusations were quite unfair, but the controversy was artificially whipped up in the media, where journalists and others tried to sensationalise what would otherwise have been a serious scientific debate. Non-scientists (and also a few scientists) often took sides according to their political views, with left-wingers holding their traditional position that all human nature is infinitely malleable, and can be improved by social engineering, while right-wingers tried to use the language of sociobiology to justify their reactionary agenda. The fact that Dawkins, Wilson and their colleagues did not always hold the views they were accused of holding tended to get lost in the heat of the debate. By the end of the century, the dust was beginning to settle, and to a large extent the sociobiologists were seen to have won the argument. The extreme left-wing view, that genetic inheritance plays no part at all in shaping human nature, was falling out of favour, and people were merely left arguing over how much, and in what ways, our genes contribute to our behaviour and mental attributes.

This section, covering all the life sciences not included in other sections, contains discoveries from a very wide variety of fields. Biologists are sometimes accused of 'reductionism', of saying that all of life is 'nothing but chemistry'. This accusation is clearly unfair—if it were true, there would be no sociobiologists, ecologists, or even physiologists, because they would all have given up those sciences long ago and taken up biochemistry instead. Living things are by far the most complex objects in the entire universe, and biologists fully realise that to study them in their entirety is beyond the reach of any one discipline.

CONDITIONED REFLEXES

Pavlov's dogs

Few branches of science excite such impassioned debate as the subject of animal behaviour, especially when it includes human behaviour in its remit. Often the scientists involved have found themselves at the centre of socio-political controversy, sometimes very much against their own will. It seems logical to infer that anything a biologist says about how animals work must have implications for ourselves, and for how we order society. George Bernard Shaw once described Charles Darwin as having 'the knack to please everybody who had an axe to grind'. This was certainly the case with Ivan Petrovich Pavlov.

Pavlov was a physiologist who, in the closing years of the nineteenth century did lengthy studies of the digestive system of dogs, and in particular to how it connected with the nervous system. This work duly earned him a Nobel Prize in 1904. However, it was only after this that he started on the work that led to his greatest discovery. He became interested in what the brain was doing in the process of learning something. Sticking closely to the subject he knew best, he looked at whether the learning process could affect the digestive tract. As everyone knows, if one puts food in one's mouth the result is a copious flow of saliva. This is called a 'reflex action'—it is apparently quite out of the control of the decision-making part of our brains. However, Pavlov found that if he rang a bell every time he put food in a dog's mouth, after a period of a few days the sound of the bell alone would be enough to cause the dog to salivate. Clearly those parts of the brain that are responsible for hearing, and more importantly for recognising the sound of a bell, have a close link to the parts that regulate reflex actions like the flow of saliva. 'Conditioning' was the word he used to describe this process of changing the stimulus that causes a reflex action.

This would have interested biologists under any circumstances, but as luck would have it, Pavlov announced his discovery at a time when Marxist thought was rapidly gaining ground among the intellectuals of Europe, and especially his native Russia. Classic Marxism held that basic human nature is entirely a product of the society in which we live (rather than the other way round), and that one could eliminate undesirable attributes such as greed and selfishness simply by creating a society in which they were not rewarded. Pavlov's discovery was music to the Marxists' ears, and in the first half of twentieth century they commonly described all human behaviour in terms of 'conditioning'. In the early years of the Soviet Union, the state paraded him as Russia's foremost scientist, even though he was openly critical of the new regime and communism in general. Today, with Marxism out of fashion, Pavlov's discovery is still seen as a major breakthrough, but nobody now suggests that it is the key to all animal and human behaviour.

I. P. Pavlov, *Huxley Lecture*, Charing Cross Hospital, London, October 1, 1906

BACTERIOPHAGES

Attacking bacteria with viruses

Ever since the late nineteenth century, it has been known that infectious diseases can be caused by two different kinds of agent: bacteria, which can only be seen with a powerful microscope, and viruses, which are much smaller still. We now know that viruses are not fully living things, but rather are 'rogue genes', encased in tiny packets of protein, which invade living cells and cause them to make more viruses. During the second decade of the twentieth century, it was found that there are even viruses that attack bacteria. These are called 'bacteriophages', often shortened to 'phages'. Today we can see them under electron microscopes. They look like tiny syringes that latch onto the outside of a bacterium and inject their DNA through its outer membrane. Within a short time, this DNA completely takes over the bacterium, causing it to manufacture hundreds more bacteriophages, which then burst out of it and go on to attack other bacteria.

There is disagreement over who was the first person to observe bacteriophages in action. The Englishman F. W. Twort reported in 1915 a disease that appeared to be infecting some bacterial cultures he was working on, which may have been caused by bacteriophages. However, the first unequivocal description was by the Frenchman François D'Herelle two years later. He was working on a bacterium that causes Shiga dysentery. One day, he extracted from a patient's excrement some bacteria that rapidly died off in his test-tube cultures, instead of multiplying in the normal way. The patient started to recover rapidly, and d'Herelle, his interest aroused, decided to experiment with the dying bacterial culture. He found that a small extract from it could kill off other cultures, and its deadliness to bacteria increased with time. Clearly something that killed bacteria was multiplying in it. It had to be a virus.

D'Herelle's discovery caused considerable excitement, because it held the promise of an effective treatment for bacterial infections. It should be possible to stop an infection dead in its tracks by giving the patient a solution containing bacteriophages. Within a few hours, they should multiply among the bacteria and wipe them out. However, this research soon ran into a number of difficulties. Scientists were soon disheartened, and shifted their attention to the new sulphonamide drugs and antibiotics, which were much easier to work with. Only in Poland and the Soviet Union did research continue on bacteriophage therapy, but that stopped in 1989 when the Eastern Bloc's economy collapsed. Today, however, several scientists are beginning to show a renewed interest in bacteriophages, as more and more bacteria become resistant to antibiotics. It may be that their time has come.

Meanwhile, bacteriophages proved a boon to geneticists. In the 1950s and 1960s, much research was done on analysing bacteriophage genes, and finding what proteins they coded for. Bacteriophages were the first organisms to have their genomes mapped in full detail, in a research programme that paved the way for modern biotechnology.

F. d'Herelle, *Comptes Rendues—Academie des Sciences*, 165, 373 (1917)

THE VERTEBRATE ORGANISER

Key to embryo development

Embryologists study one of the most obvious, but most difficult questions in all biology. How does a simple round egg grow into the complex body of an animal? Right up until the 1920s, biologists were divided over the rival doctrines of 'preformation', which held that a microscopic animal body is present in the egg right from the start, and 'epigenesis', which held that the embryo generates its complex structure as it develops. In the late nineteenth century, experimenters started working with the embryos of newts, because these are (just) big enough to be operated on, but are not encased in a shell like the eggs of birds. They tried to see if one particular part of an egg always grew into one particular part of the adult animal, or whether one could grow an entire animal from just one part of an egg. The results seemed rather conflicting.

One of these early experimenters was Hans Spemann, a senior scientist at the Kaiser Wilhelm Institute in Berlin. He used the eggs of two species, the Crested Newt and the Alpine Newt. Crested Newt embryos are white, while Alpine Newt embryos are black, so if part of an embryo from one species is replaced with a graft from an embryo of the other, one can easily see from their colour which parts of the resulting composite grow from the graft. In 1921, he gave one of his graduate students, Hilde Proescholdt, the project of grafting a part of the embryo called the 'dorsal lip' from the embryo of a Crested Newt onto the underside of an Alpine Newt embryo. The 'dorsal lip' is part of the rim of the 'blastopore', a small crescent-shaped hole that appears in the top of a developing embryo when it is a few days old, leading to a growing hollow space inside. Spemann wanted to see how an embryo grew if it had two dorsal lips, one on the top and one on the underside.

It was a difficult operation, and out of several hundred embryos, only five survived for any appreciable length of time. But these five showed a clear result. They all started to grow two backs, with two spinal chords, backbones and other tissues. The transplanted dorsal lip grew into some of the tissues of the extra back, but the extra spinal chord grew from the host embryo's tissues. Somehow the presence of the grafted dorsal lip had influenced the growth of surrounding parts of the host embryo, presumably by secreting one or more chemicals. Spemann called the dorsal lip an 'organiser'. This discovery was the final vindication of the 'epigenesis' doctrine, and the end of 'preformation'. It can fairly be called the beginning of modern embryology.

Hilde Proescholdt married another of Spemann's students, Otto Mangold. In 1924 she had a child, and published with Spemann the results of their experiment. Spemann continued research as one of the world's leading embryologists, but sadly Hilde Mangold was killed a few months later, by an exploding oil-heater in her kitchen.

H. Spemann and H. Mangold, *Wilhelm Roux's Archiv für Entwicklungsmechanik der Organismen*, 100, 599–638 (1924)

UREASE

The first enzyme to be isolated

It is no exaggeration to say that the whole of biochemistry rests on the properties and behaviour of enzymes. To the non-scientist, the word 'enzyme' may conjure up one or two associations. 'Biological' washing-powders contain synthetic enzymes which break down organic molecules in stains on our clothes; and many people probably know that we have enzymes in our digestive systems which break down the food we eat. However, to biologists enzymes have a far wider importance. Living things all work by means of very precisely-controlled chemical reactions, and it is the enzymes that provide this control. They are extremely specific catalysts, and every one of the countless chemical processes that go on in living organisms has an enzyme to control it.

The existence of enzymes has been broadly known about since the second half of the nineteenth century. By 1900 a few of them even had names, such as zymase, the enzyme in yeast that controls the conversion of sugar into alcohol, and rennin, the enzyme taken from cows' stomachs that is used to curdle milk for making cheese. However, in order to study an enzyme and its action in any detail it is first necessary to isolate a pure sample of it. The first time this was achieved for any enzyme was in 1926, when James Sumner, at Cornell University, USA, extracted pure crystals of the enzyme urease from ground-up jack beans.

Urease is not perhaps the most important of organic compounds. Sumner seems to have chosen it because it was simple to study. It is an enzyme that breaks down urea, a simple organic compound of carbon, nitrogen, hydrogen and oxygen, to make ammonia (a compound of nitrogen and hydrogen) and carbon dioxide. The beans contain urea because it is a simple molecule with nitrogen in it. Bean plants have bacteria in their roots that can extract nitrogen from the atmosphere and incorporate it into organic compounds. The plants then use these compounds as a source of nitrogen for making proteins. The extraction of ammonia from urea is just one step in this process.

Sumner's experiments were quite crude. He used a commercially-produced brand of bean meal, which he treated with acetone, a straightforward organic solvent. The urease appeared as tiny octagonal crystals. When a small quantity of these were added to a solution of urea, they instantly caused large amounts of ammonia to be generated. Sumner performed a very basic analysis on the crystals, which showed fairly clearly that they were a protein of some kind. This was a major breakthrough, because until then opinion had been divided over what kind of compound enzymes were.

We now know that enzymes are huge molecules. Urease itself is a molecule of over 100,000 atoms. Somewhere on this huge structure is the 'active site', precisely shaped to attach itself to a urea molecule and break it apart. This is the way in which all enzymes work. The exact structure of the active site is what gives an enzyme its specific ability to control just one chemical reaction.

J. Sumner, *Journal of Biological Chemistry*, 69, 435–441 (1926)

PROGESTERONE

The hormone in the Pill

It can well be argued that 'the Pill', the female oral contraceptive, is one of the most important inventions of all time, in terms of its social impact. By effectively removing the automatic link between sex and child-bearing, it has revolutionised the place of women in society, and (arguably) the role of sex in personal relationships. It is surely no coincidence that both 'feminism'—the belief that women should have the same rights and aspirations as men—and the 'permissive society'—the social acceptance of sex outside marriage as a right for men and women equally—arrived a few years after the Pill became widely available. The main ingredient of the Pill is a synthetic version of progesterone, a hormone produced in the ovaries of female mammals. Its discovery in 1929 must rank as a major turning-point in the history of science, and of society.

By 1900 it was known that a mammal's ovaries, in addition to producing egg cells, produced at least one hormone involved in the reproductive cycle, and by 1920 the first of these, oestrogen, had already been identified. However, it was clear that oestrogen was not the only hormone involved. In particular, it seemed likely that there must be a hormone that causes the uterus to enlarge in preparation for pregnancy, and keeps it enlarged if pregnancy occurs.

This hormone's existence was proved in 1929 by the

American George Washington Corner, a scholar who combined active research in physiology with a notable career as a historian of science. He was interested in the 'corpus luteum', a small mass of tissue that appears in an ovary after it has released an egg. If the animal becomes pregnant, the corpus luteum persists throughout pregnancy, but otherwise it shortly disappears. Its presence always seems to go hand-in-hand with an enlarged uterus. Corner decided that the corpus luteum must be the source of the suspected hormone, and set out to prove it. Together with Willard Allen, he prepared a corpus luteum extract, and injected it into some female rabbits. Their uteri immediately started to grow as they would during pregnancy. In a second experiment, they removed the ovaries from pregnant rabbits, an operation which normally caused them to miscarry. However, if they were given injections of corpus luteum extract, their pregnancies continued normally. Over the next few years other researchers purified and analysed the hormone in the extract, and it acquired its name, progesterone.

What Corner and Allen did not mention was that as long as a corpus luteum is present, the ovaries will not produce any more eggs. It was some twenty years later that Dr. Gregory Pincus, often referred to as the 'Father of the Pill', showed that this too is an effect of progesterone. Realising the possibilities that this discovery opened up, in 1952 he started clinical trials of progesterone as a contraceptive. The results of these trials led to the manufacture of the first contraceptive pill containing synthetic progesterone by G. D. Searle Ltd in 1956.

G. W. Corner and W. M. Allen, *American Journal of Physiology*, 88, 326–339, 340–346 (1929)

THE KREBS CYCLE

The universal power source for living tissues

We all know that we have to eat to live. But just how do our bodies extract energy from food? The simplest explanations we are given at school suggest that we simply burn our food, to produce carbon dioxide, water vapour, and lots of heat. But obviously we do no such thing. There are no flames or smoke in our bodies. The chemical reactions have to be more subtle than that.

Piecing the picture together was one of the main tasks of biochemists in the middle years of the twentieth century. The biggest piece of the jigsaw fell into place in 1937. This was the 'citric acid cycle', usually called the 'Krebs cycle' after its discoverer Hans Krebs, senior member of a family that has produced several other eminent scientists. In his experiments at Sheffield University, Krebs and his assistant W. A. Johnson minced up the muscles of freshly-killed pigeons. The pureed raw meat continued to respire as if it were still alive for at least twenty minutes, and Krebs and Johnson were able to analyse the chemical reactions going on in it. What they found was an elegant little sequence of reactions that go on in a never-ending cycle. It starts with a chemical Krebs called 'triose' (now known as 'acetyl co-A'), derived from the pigeons' food intake, which is continuously combined with a chemical called oxaloacetic acid to produce citric acid. This is then broken back down to oxaloacetic acid, which is used all over again. In the process, energy is obtained and carbon dioxide is generated.

Krebs realised he had hit on something important, and quickly sent a paper to *Nature*, the journal which has published so many of the discoveries described in this book. But on this occasion, *Nature* was scooped. The editor completely failed to see what all the excitement was about, and replied saying he had eight weeks' backlog of papers waiting to be published, so he did not need any more just now. So the prestige of announcing the great breakthrough went to an obscure Dutch journal called *Enzymologia*, which published it just two months after receiving the manuscript. Today's scientists can only dream of the time when papers appeared in print that fast—a letter sent to *Nature* usually has to wait for a few months before it is published, and other journals may take up to a year or more to get anything printed.

The discovery was bigger even than Krebs realised at the time. We now know that the cycle, or slight variations on it, occurs in very nearly all the tissues of all living things, and it must have originated very early in the history of life on Earth. In living tissues, it performs a role as important as electrical power in heavy industry—almost everything is run on it. It is not just used as a power source, but also as a starting-point for manufacturing many of the chemical compounds from which living tissues are built.

H. Krebs and W. A. Johnson, *Enzymologia*, 4, 148–156 (1937)

ATP

The energy molecule

All living things have to use energy to live and grow. Almost all of it originally comes from sunlight, which is trapped by green plants and used to power the microscopic chemical factories in their tissues that make sugars and other compounds. Other living things—animals, fungi and bacteria—then break these compounds down again and in the process liberate the energy they contain. But how is this energy carried? In factories and machines, energy is usually carried in the form of electricity or heat: what is the equivalent in living things, where there are no electric cables, and nothing is ever burnt producing flames and smoke?

The answer lies in the role of the 'phosphate group' in biochemistry. Phosphate is an 'ion'—an incomplete molecule, consisting of one atom of phosphorus and four of oxygen. It generally has to be attached to another molecule by means of a chemical bond, and this bond contains energy. All, or nearly all, the chemical reactions that take place in a living organism involve moving a phosphate group from one molecule to another, carrying energy with it. Generally, one of the molecules involved is adenosine triphosphate, always known as ATP, which will readily lose a phosphate group to become adenosine diphosphate, or ADP. In living things, converting ATP to ADP and back again is the universal way that energy is moved around.

ATP was originally discovered in 1929, by Kurt Lohmann at Heidelberg University. However, he did not realise its importance. That only slowly became apparent over the next twelve years, as biochemists started to unravel the chemical reactions that go on in living organisms. Its full role was first shown by a former colleague of Lohmann's, Fritz Lipmann, working at Cornell Medical School in New York. In a lengthy review of the role of phosphate in carrying energy, he described its circulation as being like an electric current, with the recently-discovered Krebs Cycle acting as the dynamo. As the reactions of the Krebs cycle churn out carbon dioxide, they also attach phosphate groups to organic molecules, that then go on to give them up again in any process requiring an energy source. He did not say that these organic phosphates were always ATP, but he did list fourteen known reactions in which ATP was involved. This, of course, was a tiny fraction of the true total.

There are countless different chemical substances found in living things, but the central processes that all life has in common are all based around a few molecules. For instance, ATP, as its name implies, is made of adenosine and three phosphate groups. These groups turn up widely elsewhere in living things. The DNA molecule that genes are made of also contains phosphate and adenine, a molecule very similar to adenosine. The way that so many processes all rely on the same few molecules gives biologists cause to hope that, underneath all the superficial complexity, the chemical basis of life may really be quite simple.

F. Lipmann, *Advances in Enzymology*, 1, 99–162 (1941)

BEE DANCES

Complex communication among insects

In these days when natural history programmes on television confidently claim to explain the activities of all kinds of animals, viewers may not appreciate just how difficult it is to study animal behaviour and come to any firm conclusions. It is not just a matter of watching animals and noting down what they do. Zoologists have to amass data that can be analysed mathematically, and wherever they can they try to perform experiments to test their ideas. Doing an experiment in a laboratory, like Pavlov with his dogs, is fairly straightforward. It is much harder to experiment on animals in the wild, while disturbing them as little as possible. One of the first scientists to do this successfully was Karl von Frisch, in a long-running project that ran from the 1920s to the 1940s. Throughout this catastrophic period of German history, von Frisch was quietly working with honey-bees in the seclusion of the Munich Botanical Garden. His research not only yielded detailed knowledge of one of the most remarkable of all animals, but also helped start a whole new scientific discipline: ethology, the study of animal behaviour outside laboratory conditions.

The question von Frisch asked was: how do bees forage so efficiently? They cannot all just be searching for flowers at random. The answer he found was that when a bee finds a clump of flowers with plentiful nectar and pollen, it can fly back to the hive and tell all the other bees where it is. It does this by 'dancing'—performing little stereotyped movements on the surface of a honeycomb, while other bees cluster round it. There are two kinds of dance: the 'round dance', which simply means 'there are flowers with nectar close to the hive', and the 'waggle dance', which actually shows which direction the bees must fly in order to find more distant flowers (*see* Plate 15). It is little short of astonishing that an insect, whose brain is so small that it is almost certainly unaware of its own existence, can communicate with its fellows in this way.

Von Frisch's method was straightforward, but slow. Firstly, he and his assistants set up an 'observation hive'—a beehive with glass-panelled sides, inside a hut where it could be watched, but with its entrance hole leading to the world outside. Next, he set up 'artificial flowers', consisting of watch-glasses containing sugar syrup resting on pieces of brightly-coloured paper, for the bees to find. When bees visited the artificial flowers, he marked them with spots of paint, so that they could be recognised when they returned to the hive. He soon saw the bees dancing, and then over several years of painstaking experimentation he deciphered the dances by observing how they changed if the artificial flowers were moved from one place to another.

There is no Nobel Prize for Biology, but in 1973 von Frisch, together with two other pioneering ethologists Niko Tinbergen and Konrad Lorenz, was awarded a Nobel Prize for physiology and medicine, in recognition of his work in founding a new science.

K. von Frisch, *Osterreicher Zoologischer Zeitschrift*, 1, 1–48 (1946)

EVOLUTION DRIVEN BY COMPETITION

Darwin vindicated

The theory of evolution has been slow to mature. The idea originated at the end of the eighteenth century, but it only became a truly scientific theory in the 1850s, when the two British biologists Charles Darwin and Alfred Russell Wallace independently suggested a mechanism for it: natural selection. According to Darwin and Wallace, the motor that drives evolution is the never-ending struggle for existence, in which all living things must adapt or perish. This theory is arguably the most radical thing that science has ever produced. However, for nearly a hundred years scientists only partly agreed with it. In the first half of the twentieth century, most biologists accepted the concept of evolution, but did not believe that it was being driven by natural selection. Rather, they saw natural selection as something that prevented change, by weeding out any organisms that were different from their parents and thus (presumably) less well-adapted to their environment.

The first sign that Darwin and Wallace could be right came in the 1920s, when the theorist R. A. Fisher first put evolution theory in mathematical terms, and showed that it could work. However, in those days very few biologists were mathematicians, and the only thing that would convince them was solid evidence. This was finally produced in 1947 by a British ornithologist called David Lack. He had done the logical thing, by starting exactly where Darwin had left off. Just before the Second World War he went to the Galapagos Islands to study the birds that had so impressed Darwin a hundred years earlier: the Galapagos finches. There are fourteen species of finch on the Galapagos, and they all look so broadly similar that no-one doubts they are all descended from a common ancestor. They are rather dull-looking birds, coloured black and brown. They differ in size, particularly in the size of their beaks, and in the food they eat. Darwin had only stopped in the Galapagos for a few days, and had no time to do more than collect some specimens. Lack stayed for several months, and studied the finches' ecology intensively. He found that wherever two species with similar-sized beaks were found on two different islands, they would live in similar habitats and have similar diets. But wherever the same two species were found on the same island together, they inhabited different habitats and had different diets.

The outbreak of the Second World War forced Lack to return to Britain. He spent the next seven years poring over his data, and was forced to one conclusion. The birds must be actively competing with each other for resources, and there was no room for losers. A species may not share its living space with another whose beak and diet are similar to its own. From this it was logical to infer competition had driven the Galapagos finches to evolve into different species with different beaks and diets. Lack published his studies in 1947, and since then few biologists have looked for any other explanation for evolution. Darwin was right after all.

D. Lack, *Darwin's Finches*. Cambridge University Press (1947)

HOW NERVES WORK

The action potential

All but the simplest of animals have a nervous system, a network of cells used for passing information around the animal's body. In most animals there is a central part or 'brain', where huge numbers of nerve cells connect with each other, and radiating out from it is a system of nerves which serve to bring in information from the animal's senses, and also to convey instructions for behaviour out to other parts of the body. Nerves consist of bundles of ultra-thin fluid-filled tubes called 'axons', each growing out from a single cell. In a working nerve, signals are passing along these axons. But what is the precise nature of the signal?

The first clue came as long ago as the 1790s, when the pioneer of electricity Luigi Galvani showed that the dissected legs of a frog would kick if a pulse of electric current were applied to the nerves leading into them. It would seem that nerve signals are of an electrical nature. This discovery provided the inspiration for Mary Shelley's classic novel *Frankenstein*, but further understanding did not come until the mid-twentieth century. In the 1930s Alan Hodgkin of Cambridge University, also working on the nerves of frogs, showed that the outer membrane of an axon acts like an electric capacitor. Its outside surroundings have a marked positive electrical charge compared with the fluid inside. However, when a nerve signal (or 'action potential') passes along an axon this charge is momentarily reversed. The action potential consists of a short region of reversed charge, travelling along the axon at high speed.

Hodgkin's research was interrupted by the Second World War, but in the late 1940s he teamed up with Andrew Huxley (brother of the novelist Aldous Huxley), and started working on the axons of squids. Most axons are microscopically thin, but squids also have two 'giant axons', about a millimetre in diameter, big enough for electrodes to be inserted right into them. In a series of four papers published in 1952, Hodgkin and Huxley showed that the mechanism of the nerve impulse worked with positively-charged atoms ('ions') of sodium and potassium. When an axon is resting, it maintains a considerable excess of sodium ions outside it, and a similar excess of potassium ions inside. As an action potential sweeps along an axon, its outer membrane becomes 'depolarised'—it suddenly lets the sodium ions flow through it into the axon, reversing the electrical charge. A split second later, potassium ions flow the other way, and the original electric charge is restored. This reversal of the electrical charge triggers further depolarisation in adjacent regions of the axon, resulting in the action potential sweeping down it like a wave.

This discovery gave neurology a firm basis in chemistry. Further progress in this field has proved of immense medical value, in the development of psychoactive drugs, and in the understanding of some disease conditions including Parkinson's disease and alcoholism. Its use in developing 'nerve gases' for the military, however, is less creditable.

A. L. Hodgkin et al, *Journal of Physiology*, 116, 424–448; A. L. Hodgkin and A. F. Huxley, *Journal of Physiology*, 116, 449–472; 473–496; 497–506 (1952).

THE CHEMICAL COMPOSITION OF INSULIN

The first protein molecule to have its amino acid sequence analysed

All of life is built on two types of molecules: nucleic acids and proteins. The nucleic acids are the information-carriers, the molecules that genes are made of. Proteins, meanwhile, form the machinery of a living organism. The first step in understanding this machinery must be to find the composition and structure of the protein molecules. These are made of smaller units called amino acids. There are twenty different kinds of amino acid found in proteins, and the exact pattern in which they are linked together in a protein molecule determines its overall shape and functioning. The first protein to have its composition analysed in full was insulin, the hormone which controls the rate at which our bodies assimilate sugar. This was achieved by the Cambridge biochemist Frederick Sanger, one of the very few people ever to win two Nobel Prizes.

Sanger started work on the project in 1944, and it took him and his associates twelve years in all. He chose insulin mainly because it was commercially available in highly purified form, for use in treating diabetes. It also had the advantage of being a fairly small molecule, as proteins go. He already knew that it consists of two convoluted chains, in which each link is an amino acid. These chains turned out to be linked by four 'bridges'. Sanger's task was firstly to separate the two chains from each other, and then to find the sequence of amino acids in each chain—assuming that there was one. Not all biochemists believed this; some suspected that while the overall composition of all insulin molecules was the same, the exact sequence of the amino acids could vary at random from one individual molecule to the next. Sanger showed that this was not the case. All insulin molecules are indeed the same, and the exact sequence of amino acids is very specific. His method of working out this sequence consisted of breaking up the chains into many short overlapping fragments, analysing these fragments' compositions, and then working out how they must overlap. It was a task beset with difficulties, but eventually in 1953 he was able to publish a complete sequence for each chain, showing twenty-one amino acids in the first chain and thirty in the second. The analysis was finally completed in 1955, when he showed the positions of six 'amide' groups, consisting of one nitrogen and two hydrogen atoms each, attached to amino acids on each chain.

Sanger had to develop his techniques as he went along, finding out much about the basic composition of proteins in the process. Today, protein sequencing is a routine business, done largely with automated laboratory equipment and purpose-built computer software. Most protein molecules are far bigger than the insulin molecule, and biochemists no longer write the sequences out in full when they publish their results. Rather, they upload them into huge online databanks, from which anyone who needs them can print them out. However, the methods used to generate this data are still essentially those developed by Sanger over fifty years ago.

F. Sanger and E. O. P. Thompson, *Biochemical Journal*, 53, 353–366; 366–374 (1953)

THE STRUCTURE OF VITAMIN B12

Full analysis of a biological molecule

The chemical compounds from which living things are made—the proteins, nucleic acids, and all the rest—are by far the most complex in existence, with individual molecules containing thousands or even millions of atoms. Biologists have always known that if they really want to know how living things work, they must analyse these molecules, to find not only their chemical composition, but also their exact structure and shape. However, until the middle years of the twentieth century the task seemed too difficult to contemplate. Working out a molecule's three-dimensional structure—the positions of all its constituent atoms—is done by X-ray diffraction crystallography, a technique developed in the early years of the twentieth century. It involves shining a beam of X-rays through a very pure crystal of a substance. The crystal causes the beam to split up into a complex pattern, which can then be analysed mathematically to show the positions of individual atoms in a molecule. It is not too difficult to analyse simple compounds by this method, but applying it to giant biological molecules is another matter entirely.

However, the scale of the task did not daunt Dorothy Hodgkin (née Crowfoot), a young biochemist at Oxford University. Starting in the late 1930s, her first success came with penicillin, a molecule containing 39 atoms. Then, after some unsuccessful years trying to solve the structure of insulin (she finally succeeded in 1969), she and her team turned their attention to Vitamin B12, the vitamin required to prevent pernicious anaemia, whose molecule contains nearly two hundred atoms. In 1955 she was able to publish a preliminary analysis of its structure, with a more definitive version following a year later.

Vitamin B12 is still not a very large molecule by biological standards, but it was by far the biggest to have been analysed at that time. One of the original inventors of X-ray analysis, W. L. Bragg, later described Hodgkin's achievement as 'breaking the sound barrier' in its field. Two factors had contributed to making it possible. Firstly, her team had access to some of the earliest mainframe computers, enabling them to do calculations that had previously been unthinkable. Secondly, unlike most biological molecules, Vitamin B12 contains an atom of a heavy metal, cobalt. The presence of heavy atoms greatly helps in X-ray diffraction analysis. By working on a molecule that already contained a heavy atom, Hodgkin had an easier job than scientists such as Max Perutz and John Kendrew at Cambridge, who were starting to analyse protein molecules by adding heavy atoms to them artificially.

Dorothy Hodgkin is one of only four women ever to have received a Nobel Prize for science. She was a cheerful and much-liked character, immensely respected by her many pupils and colleagues. In addition to her scientific research, she was also noted for her pacifist work, especially as one of the organisers of the influential Pugwash conferences, where scientists met to discuss ways of reducing the risk of nuclear war. She died in 1994.

D. C. Hodgkin et al, *Nature*, 178, 64–66 (1956)

THE CALVIN CYCLE IN PHOTOSYNTHESIS

How green plants provide us with food and oxygen

As everyone knows, green plants take up carbon dioxide from the air and water from the soil, and turn them into carbon compounds and oxygen, using sunlight as an energy source. All other living things, including ourselves, rely on this process as the ultimate source of the food we eat and the oxygen we breathe. It is called 'photosynthesis', and its broad outline has been known ever since the late eighteenth century. But what are the exact chemical reactions involved? This was the question addressed by Melvin Calvin, a biochemist at the University of California at Berkeley, together with his colleagues James Bassham and Andrew Benson. Between them they worked out the set of chemical reactions at the heart of photosynthesis, ever since called the 'Calvin Cycle'.

Their method involved carbon-14, an isotope of carbon that is radioactive. Calvin's idea (since used very widely in biochemistry) was to use the radiation coming from the carbon-14 to track where it had got to. The plant he studied was one of the simplest—an aquatic microbe called *Chlorella pyrenoidosa*, which he grew in culture bottles under a bright light, with carbon dioxide bubbling through. His technique was suddenly to introduce a momentary pulse of carbon dioxide made with carbon-14, and then a few seconds later to kill the *Chlorella* and analyse the carbon compounds it contained. Any compounds that were radioactive must be products of photosynthesis, made from the radioactive carbon dioxide to which the *Chlorella* had just been exposed.

Calvin and his team spent some ten years working on this project, and identified a number of compounds that must be involved. It appeared that the chemical reactions followed a cycle, in which carbon dioxide was added at one point, and the sugar glucose came off at another point. But not all of the evidence seemed to fit together. Then in 1958, Calvin had a 'eureka moment'. The story is best told in his own words: 'One day I was sitting in the car, while my wife was on an errand. While sitting at the wheel of the car, the recognition of the missing compound occurred just like that—quite suddenly. Suddenly also in a matter of seconds the complete cyclic character of the path of carbon became apparent to me . . . [It] all occurred to me in the matter of thirty seconds.' The missing link in the cycle that Calvin had just worked out was a compound called phosphoglyceric acid. It is made from a compound Calvin already knew about called ribulose diphosphate, by the addition of carbon dioxide. The end-product of photosynthesis, glucose, is generated in a chain of reactions that converts the phosphoglyceric acid back to ribulose diphosphate, to be used again.

Calvin firmly believed in promoting links between scientists with different specialisms. When he became head of a large laboratory at Berkeley, he designed an open-plan building for it, so that all the researchers could have easy access to each other. This sound principle has only recently started to become fashionable in laboratories elsewhere.

M. Calvin, *Science*, 135, 879–889 (1962)

MONOCLONAL ANTIBODIES

How to mass-produce an antibody to order

One of the more extraordinary capabilities of the human body is its 'immune system', the mechanism by which we recognise foreign organisms such as bacteria, or tissues transplanted from another organism. What the body does seems quite incredible: it can immediately identify any foreign protein, and rapidly produce an 'antibody'—another protein molecule that will bind to that protein and no other. No analyst's laboratory could ever be anywhere near so versatile. So when, in 1975, a way was found to harness the system and produce a single antibody in the laboratory in unlimited quantities, it caused major excitement.

Antibodies are produced by certain white blood cells, called B lymphocytes. Any single B lymphocyte only generates one antibody, but new lymphocytes, making new antibodies, are being produced all the time. If one's body encounters a foreign protein, belonging to some invading bacteria for instance, it somehow monitors what happens to it. As soon as an antibody is found that binds to it, lots more B lymphocytes producing that particular antibody are produced, and the invading bacteria (or whatever) are soon smothered with antibody and killed.

As soon as biologists began to understand even broadly how the mechanism worked, they dreamed of using it for their own ends. A mass-produced antibody could be used as a totally specific drug for fighting a disease. Alternatively, it could be used as a research tool, for homing in on a particular protein—a capability that has many possible applications in medicine and biology. But to mass-produce an antibody, it seemed one had to mass-produce the lymphocytes that generate it, and that just could not be done. Lymphocytes are produced in bone marrow, but unlike most other cells in our bodies, they do not reproduce themselves, so they cannot be grown in culture in a test-tube.

This was the problem solved by the German Georges Kohler and the Argentinian Cesar Milstein, working at the MRC Molecular Biology Laboratory at Cambridge. Their method was to fuse B lymphocytes with a kind of cancer cell called a 'myeloma', which is potentially immortal—it can go on reproducing itself indefinitely. They called the resulting hybrid cell a 'hybridoma', and it combined the antibody-producing ability of the lymphocytes with the immortality of the myeloma. They then screened the hybridoma cells, and cloned those that were producing the desired antibody. A culture of these cloned cells produced the 'monoclonal antibody'—one specific antibody, in limitless amounts.

Monoclonal antibodies were soon being widely used for scientific research, but their use for medical purposes has proved problematic, and as yet only a handful are on the market as drugs. The problem is that the hybridomas are all derived from mice rather than humans. The human body can recognise mouse antibodies as foreign, and will attack them with antibodies of its own. However, research is now being directed at ways of 'humanising' mouse antibodies so that the human body will accept them, and there is every hope that this problem will shortly be solved.

G. Kohler and C. Milstein, *Nature,* 256, 495–497 (1975)

GENETICS

If one thinks of the first half of the twentieth century as being the great age of physics, then the second half surely belongs to genetics. Especially in the late 1940s and 1950s, genetics was widely seen as the place to be, where the most exciting discoveries were waiting to be made. Broadly speaking, this excitement followed the discovery that genes are made of DNA. A technique for finding the structure of the DNA molecule—X-ray diffraction analysis—was available, and it was clear that the next few years were going to see major breakthroughs. These would in turn lead to progress on one of science's biggest mysteries: how living things work at a molecular level, the chemical reactions that mark life as being different from non-life. Without genetics, the rest of biochemistry could tell no more than half the story.

However, the achievements of the previous half-century, though they came more slowly, were no less important. In 1900, genetics was still effectively a closed book. The foundations of the science had been laid as long ago as the 1860s by the Austrian monk Gregor Mendel, but his work had passed almost unnoticed at the time, and was only rediscovered in the early 1900s. This rediscovery paved the way for a whole string of advances, as geneticists arrived at a working definition of what a 'gene' really is; pinpointed where they are located within a living organism; and worked out how they are passed on from one generation to the next.

In the early years, these discoveries helped to give evolution theory a sound basis. Few people today realise that when Darwin first propounded his theory in the 1850s, it was manifestly incomplete. He did not know that there are discrete 'genes' that may be either expressed or suppressed. Rather, he believed, along with most other people, that an organism inherits characteristics from its parents in an continuous blend. This would have meant that if a new, mutant characteristic arose in an individual, it would not be fully passed on to any of that individual's offspring, but would rather become diluted out of existence within a few generations. It is hard to see how evolution could ever occur if this were the case, and Darwin was well aware of this problem with his theory. The early discoveries of T. H. Morgan and his colleagues provided the solution, enabling biologists to develop the 'neo-Darwinist' evolution theory that is accepted today.

Right from the start, the world at large was interested in genetics. People realised at once that this was a science with implications for society as a whole. In the first half of the twentieth century, however, the chief impact came not from mainstream geneticists, but from people who misunderstood the subject. Their effect was entirely pernicious. First of all, there was 'eugenics', popular in America, and also practised as official policy in Nazi Germany. The notion that one can improve the 'fitness' of a human population firstly requires that everyone should agree on what constitutes a 'fit' person (to the eugenicists, it usually seemed to be related to earning capacity). More seriously, it assumes that 'fitness' is a single characteristic that can be inherited. This is very far from the case. Our genes all interact with each other, and also with environmental factors, in a myriad of ways. Forcibly preventing the least 'fit' individuals from having children will never rid society of 'unfit' members, however one might choose to define the term. Eugenics merely curtails peoples' freedom to raise families, to no good purpose.

The eugenics movement was a product of capitalist society, in which it is generally assumed that social inequality is a fact of life. Eugenics was touted as the 'solution' to the 'problem' of society's least competitive members. In the Soviet Union, meanwhile, communism spawned Trofim Lysenko. A charlatan who found favour with Stalin, Lysenko insisted that the Western geneticists were wrong, and that living things could pass on in their genes characteristics that they acquired during their own lifetimes. This notion was popular with Marxists, because if it were true, it would be easier for a communist society to produce the 'New Men' prophesied by Marx, devoid of selfish tendencies. Lysenko rose to become Director of the Institute of Genetics of the Soviet Academy of Sciences, and from this position he enforced this orthodoxy on genuine scientists, thus bringing to a halt all serious biological research in the Soviet Union. Scientists who opposed his views were hounded out of their jobs, and sometimes even imprisoned and shot. Meanwhile, attempts to apply his doctrines to agriculture contributed to the disastrous famines of the Stalin era. Science and wishful thinking do not mix.

Today these fads have become outmoded. However, genetics has not ceased to be controversial, indeed quite the opposite. From the 1960s onwards, geneticists became increasingly interested in the possibility of putting their newly acquired knowledge to practical use, in manipulating genetic material. The ethical implications were obvious. Firstly, there was the simple issue of safety. Could scientists inadvertently (or even deliberately) create a new and devastating disease by changing the genome of a bacterium or a virus? But beyond this, there are more general and philosophical questions. To what extent is it acceptable for scientists to change living things, including possibly human beings, by manipulating their genes? Just what sort of a brave new world are we about to find ourselves in? It should be said that scientists are not all as uncritical in their support of genetic engineering as their opponents portray them. Nonetheless, it is probably fair to say that they tend to find the future possibilities more exciting, and less frightening, than non-scientists do. The onus is on the scientists to convince the world that genetic engineering really is as desirable as they think.

ONE GENE—ONE ENZYME

The true definition of a gene

By the year 2000, the science of genetics was seen to be one of the twentieth century's most important legacies for the future, probably more significant even than nuclear physics or information technology. Genes were discussed endlessly in books, newspaper articles, and conversations, often by people who seemed only vaguely to understand what a gene is. However, to the geneticists themselves, it was vital to know exactly what a gene is, in order to begin study on how it might work.

The original concept of the 'gene' comes from the experiments of Gregor Mendel, and particularly from the biologists who re-discovered his work in the early twentieth century. To them, a gene was a 'unit of heredity', a piece of information that determines one discrete characteristic in an organism. For instance, Mendel had found that his peas could either be coloured green or yellow, and one discrete piece of information passed down from one generation to the next would determine whether a pea plant produced green peas or yellow ones. So there was one 'gene' for pea colour. This description tells us broadly what a gene does. But it still does not tell us what a gene really is. Today, biology students are taught that the true nature of a gene was first described by two Americans, George Beadle and Edward Tatum, in 1941: 'genes . . . regulate specific reactions . . . by determining the specificities of enzymes' (usually quoted as 'one gene—one enzyme'). All the chemical reactions in a living organism are controlled by enzymes, protein molecules that act as very precise catalysts. The molecular design of an enzyme is crucial to its working, and a gene is the specification for that design.

However, Beadle and Tatum were not the first to demonstrate the link between genes and enzymes. That distinction goes to Archibald Garrod, a British physiologist who made essentially the same discovery over thirty years earlier. In 1909, he gave a series of lectures to the Royal College of Physicians, describing his studies on a rare condition called 'alkaptonuria', in which a person's urine turns black when exposed to light. In the first lecture, he described how alkaptonuria was an inherited condition, and showed that it was 'a rare recessive character in the Mendelian sense'—in other words, it is caused by a single mutant gene. Then in the third lecture, he described a whole series of experiments by which he established that alkaptonuria is caused by the failure of one particular enzyme involved in the digestion of two common organic compounds, tyrosine and phenylalanine. The breakdown is left half-finished, leaving an unwanted by-product called homogentistic acid. This is the compound that turns black in daylight.

Genetics was not Garrod's main interest, which may explain why he did not bother to bring his two discoveries together and announce that he had nailed down the true nature of a gene. He clearly realised it, but he seems not to have considered it important. As a result, his name is largely forgotten, while Beadle and Tatum received a Nobel Prize.

A. E. Garrod, *Inborn Errors of Metabolism*, Oxford University Press (1909).

GENES ON CHROMOSOMES

The birth of a new science

As everyone learns at school, the science of genetics was founded in the nineteenth century by a German monk called Gregor Mendel, whose work was promptly forgotten for thirty years. It was rediscovered around the year 1900 by scientists who had started work on the heredity in plants, only to find that someone had been there before. Mendel's main discovery had been that a living thing's characteristics are inherited as a combination of discrete elements, which we now call genes. Genes occurred in pairs. Some were mutants, but these were usually 'recessive'—their effect would be masked unless both genes in a pair were mutant. The first questions facing biologists were: Where are these genes located? What are they made of?

In the 1900s, some researchers began to suspect that the genes could be located in the chromosomes, small thread-like bodies found in the nucleus of each living cell in every organism's body. A good deal was being discovered about chromosomes; how they occurred in pairs, except in egg and sperm cells, which only had one chromosome from each pair; and how one chromosome was the 'X' or 'sex chromosome'. In most organisms, females are 'XX'—they have two identical X chromosomes, while males are usually 'XY'—one chromosome of the pair is missing a large piece. These facts suggested that the chromosomes could have something to do with heredity. Could

Mendel's paired genes be located on the paired chromosomes? It seemed a reasonable idea, but not everyone was convinced. One doubter was Thomas Hunt Morgan, Professor of Zoology at Columbia University in America. He had chosen to use the little fruit fly *Drosophila melanogaster*, that abounds in greengrocers' shops, for his experiments (*see* Plate 14). It was easy to breed in the laboratory, and a new generation of flies emerged every twelve days or so. It soon became every geneticist's favourite animal, and Morgan was soon to make the discovery that can fairly be said to mark the beginning of modern genetics.

One day in 1909, there appeared among Morgan's flies a strange specimen; a male fly whose eyes were white, instead of the normal red. Morgan soon found that the white eyes were caused by a 'recessive' gene, but one with a difference: only male flies ever had white eyes. Further study forced him to the conclusion that the gene must be located on the X chromosome, and specifically on that part of the X chromosome that is missing in the male 'Y' chromosome. So when a male fly has one copy of the white-eye gene, there is no red-eye copy on another chromosome to mask its effect. His surprise shows through in the wording of his article in *Science* magazine, which does not even contain the word 'chromosome', although his use of the terms 'X' and 'XX' makes it quite clear what he is talking about. Contrary to his previous beliefs, chromosomes are indeed the places where the genes reside, and he could even point to a region of a particular chromosome and say: 'The gene controlling eye colour is located there'.

T. H. Morgan, *Science*, 32, 120–122 (1910)

GENES ARE MADE OF DNA

The 'molecule of life'

In the early decades of the twentieth century it was clearly established that an organism's genes are located on its chromosomes. But what are they made of? Oddly enough, the suggestion that they might be made of DNA (deoxyribonucleic acid, to give its full name) was made as long ago as the 1880s, but no-one took the idea seriously at the time. Chromosomes were known to be made of protein and DNA, but people assumed that the genes were in the protein part. DNA just seemed such a boring molecule, made up of only six component parts—ribose, phosphate, and the four 'bases' guanine, cytosine, adenine and thymine. How could such a straightforward chemical carry all the information needed to build a living organism? Most likely, people said, it provided some kind of structural scaffolding in the chromosomes, on which genes consisting of protein molecules were arranged.

As is often the case in scientific research, the truth became apparent from several different pieces of evidence. The first and most important of these was produced by a medical researcher in America called Oswald Avery, who in 1944 was working on *Diplococcus pneumoniae*, the bacterium that causes pneumonia. This was research of the highest importance, because before the arrival of antibiotics pneumonia was the biggest single cause of death in the western world. Avery had been working on it for many years, and was about to retire. He was trying to find a way to counter the bacterium's virulence. He knew that the bacterium came in two types: the normal, deadly, type with a smooth outer covering, and a mutant 'rough' type, which was very much less virulent—if it were present on its own. However, it had been known for some eighteen years that 'rough' bacteria only had to come into contact with 'smooth' bacteria to regain their 'smoothness' and deadliness. Most strangely *the 'smooth' bacteria had this transforming effect even if they were dead.* The rough bacteria could evidently make themselves smooth by absorbing a substance from the dead smooth bacteria, and they then passed their acquired 'smoothness' on to their offspring. Almost by definition, the thing they had absorbed must be the gene for smoothness. Avery, together with a team of co-workers, set to work to find out what it was made of.

This was an enormous task. The dead bacteria contained proteins, fats, carbohydrates, nucleic acids, including DNA, and much else. Avery's approach was to remove or destroy each of these components of the mix one by one, until he found the culprit. He soon ruled out proteins. He treated the dead smooth bacteria with powerful enzymes that would break down any protein, and the resultant smashed-up mess still had its transforming ability. Enzymes that broke up fats or carbohydrates were equally ineffective. Finally, he tried an enzyme that broke down DNA. Immediately, the dead bacteria were rendered harmless.

Avery was a modest man, and the paper announcing his results did not trumpet that 'genes are made of DNA'. But it did not have to. Everyone saw that his results could only mean that one thing.

O. T. Avery et al, *Journal of Experimental Medicine*, 79, 137–157 (1944)

GENE RECOMBINATION IN BACTERIA

Sex, of a sort

The progress of science is often advanced by pure luck. One notable happy accident was the decision by geneticists to concentrate much of their research on the bacterium *Escherichia coli*, simply because it is extremely common, easy to breed in a culture bottle, and (usually) completely harmless. Only later was it found that *E. coli* has one really useful trick that is shared by relatively few other kinds of bacteria: it indulges in gene recombination, or 'bacterial sex'. This discovery has been put to good use, enabling geneticists to construct maps of the *E. coli* genome, and to work out how the different genes interact with each other.

Joshua Lederberg was a noted geneticist at Yale University. In 1946 he was working with his student E. L. Tatum, studying the ability of bacteria to take up and use each others' DNA. This had been known about for some five years, ever since Avery first noted it and deduced that DNA is the molecule that genes are made of. For most kinds of bacteria, we still do not really know how they do it, but probably they are absorbing DNA leaking out from dead individuals. But Lederberg and Tatum found that *E. coli* was doing something different: most of the bacteria did not take up each other's DNA, but a few were taking it on board piecemeal, so they acquired various combinations of each others' genes. It was later found that the bacteria had to come

1 Einstein at the blackboard in the 1930s.

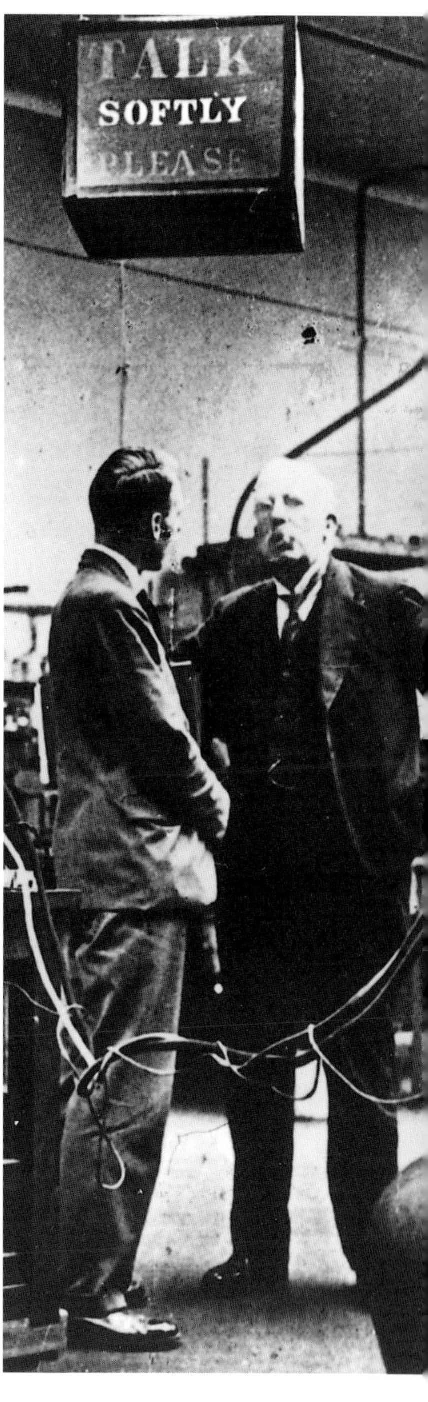

2 Rutherford (*right*) in his laboratory at Cambridge in the 1930s (*see* Nuclear Fusion).

Für "Science Illustrated"
Jan. 1946

Das Gesetz von der Aequivalenz von Masse und Energie ($E = mc^2$)

In der vor-relativistischen Physik gab es zwei voneinander unabhängige Erhaltung bezw. Bilanzgesetze, die strenge Gültigkeit beanspruchten, nämlich

1) den Satz von der Erhaltung der Energie
2) den Satz von der Erhaltung der Masse.

Der Satz von der Erhaltung der Energie, welcher schon im 17. Jahrhundert von Leibnitz in seiner vollen Allgemeinheit als gültig vermutet wurde, entwickelte sich im 19. Jahrhundert wesentlich als eine Folge eines Satzes der Mechanik. Man betrachtete ein Pendel, dessen Masse zwischen den Punkten A und B hin und her schwingt. In A (und B) verschwindet die Geschwindigkeit v, und die Masse m steht um h höher als im tiefsten Punkte C der Bahn. In C ist diese Hubhöhe verloren gegangen; dafür aber hat die Masse hier eine Geschwindigkeit v. Es ist, wie wenn sich Hubhöhe in Geschwindigkeit und umgekehrt restlos verwandeln könnten. Die exakte Beziehung ist

h A C B

$$mgh = \frac{m}{2} v^2,$$

wobei g die Beschleunigung der Erdschwere bedeutet. Das Interessante dabei ist, dass diese Beziehung unabhängig ist von der Länge des Pendels und überhaupt von der Form der Bahn in welcher die Masse geführt wird. Interpretation: Es gibt ein etwas (nämlich die Energie) die während des Vorgangs erhalten bleibt. In A ist die Energie eine Energie der Lage oder „potentielle Energie", in C eine Energie der Bewegung oder „kinetische Energie". Wenn diese Auffassung das Wesen der Sache richtig erfasst, so muss die Summe

$$mgh + m\frac{v^2}{2}$$

auch für alle Zwischenlagen denselben Wert haben, wenn man mit h die Höhe über C und mit v die Geschwindigkeit in einem beliebigen Punkte der Bahn bezeichnet. Dies verhält sich in der That so. Die Verallgemeinerung

3 (*above*) The Law of Equivalence of Mass and Energy ($E = mc^2$). Einstein manuscript, dated 1946.

4 (*left*) Physics then. Rutherford's laboratory at McGill University, Canada, in the early 1900s.

5 (*right*) Physics now. One of the four particle detectors attached to the Large Electron Proton collider at CERN.

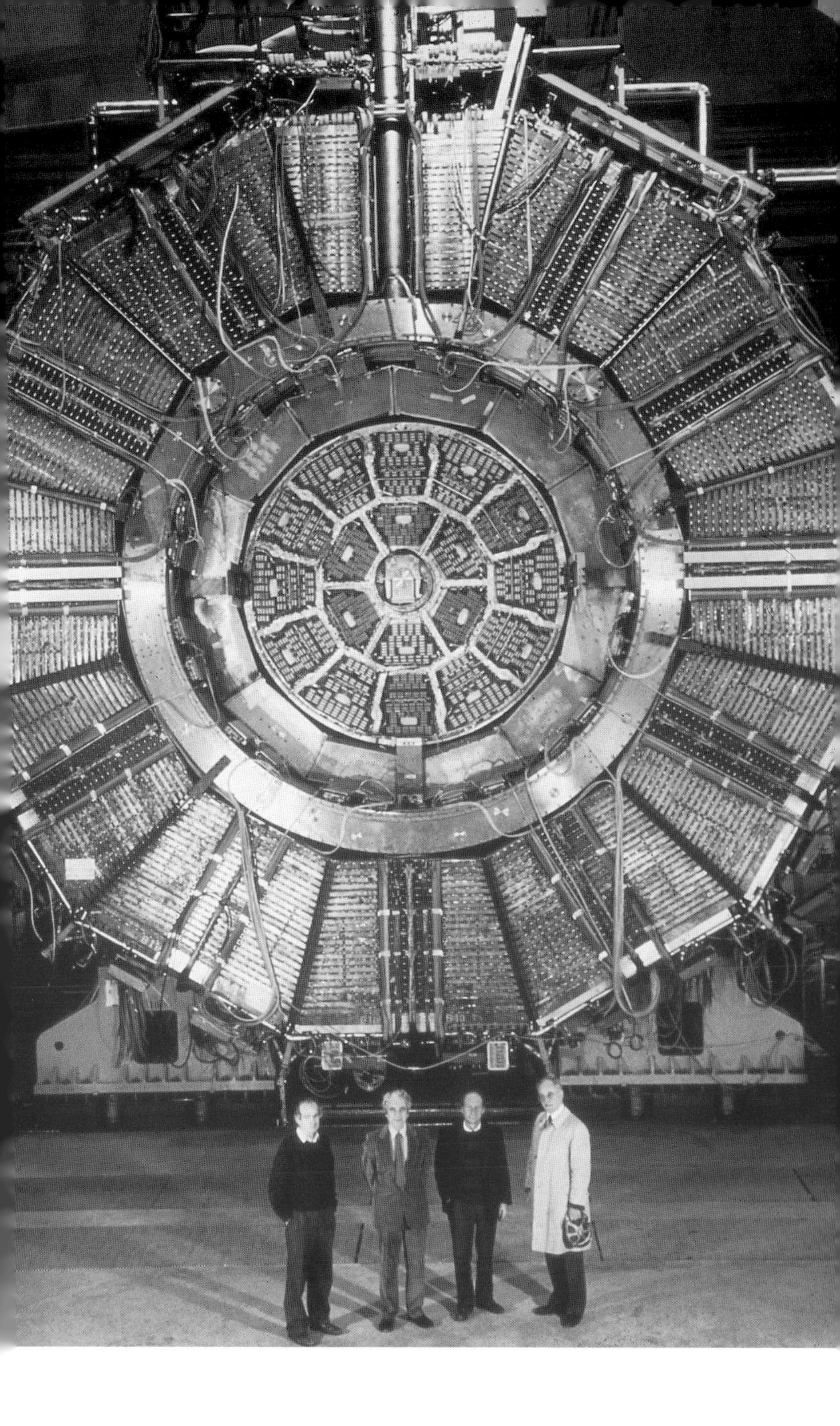

6 (*right*) Harrison Schmidt, the only qualified geologist to visit the Moon, on the Apollo 17 mission in 1972 (*see* The Age of the Solar System).

7 (*below*) The Andromeda Nebula, shown by Edwin Hubble to be an entire separate galaxy over a million light-years away.

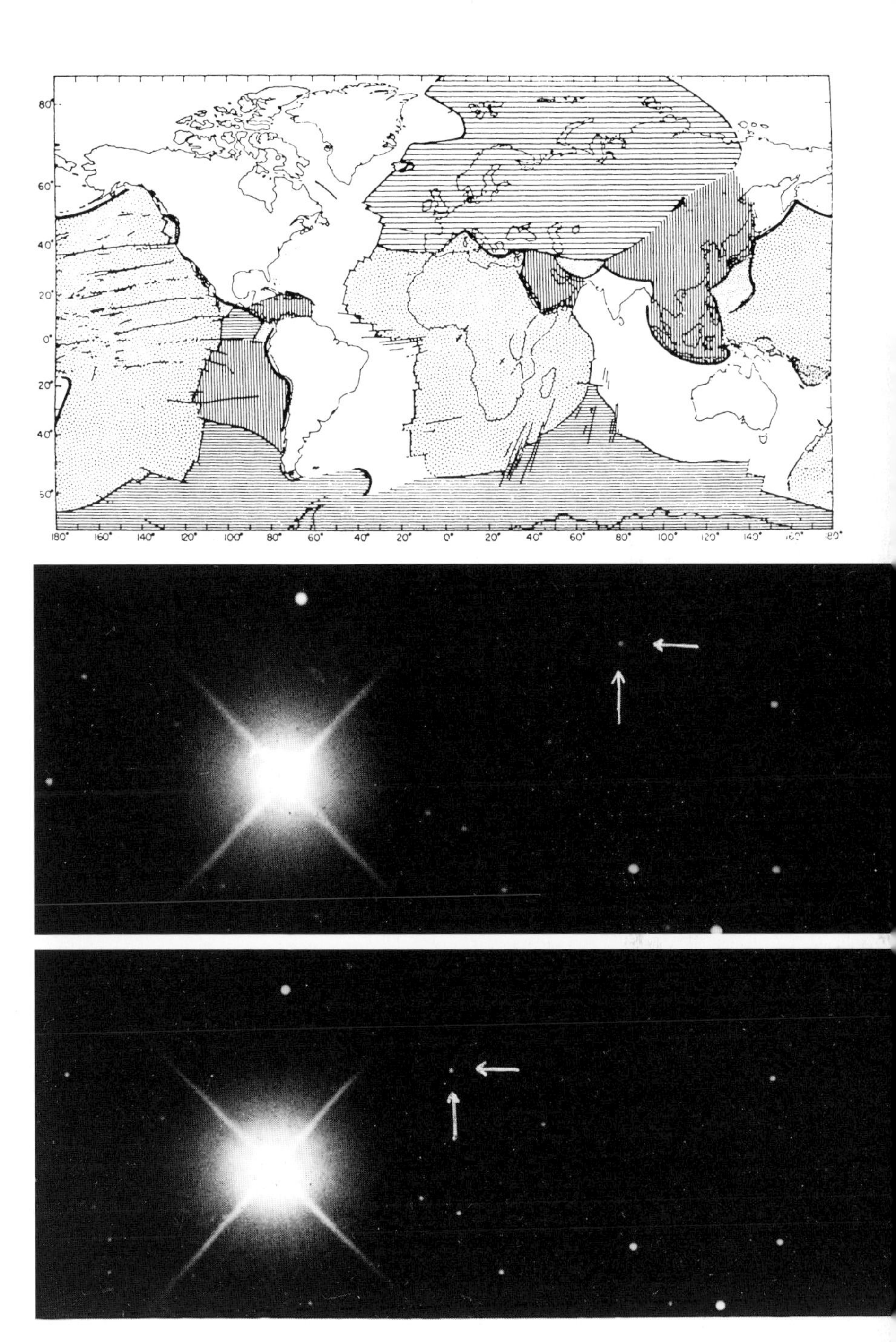

8 (*top*) Plate Tectonics. William Morgan's 1968 world map, showing the different geological 'plates'.

9 (above) Pluto. Two photographs of stars, taken three days apart in March 1930. Pluto (*marked with arrows*) is the 'star' that has moved.

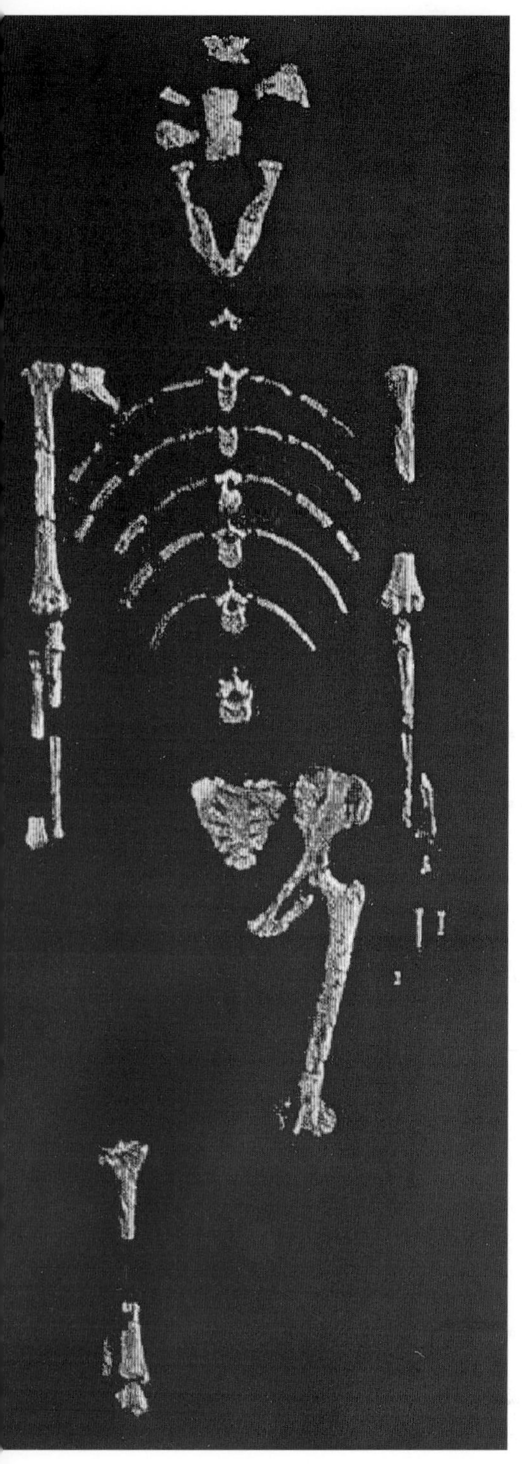

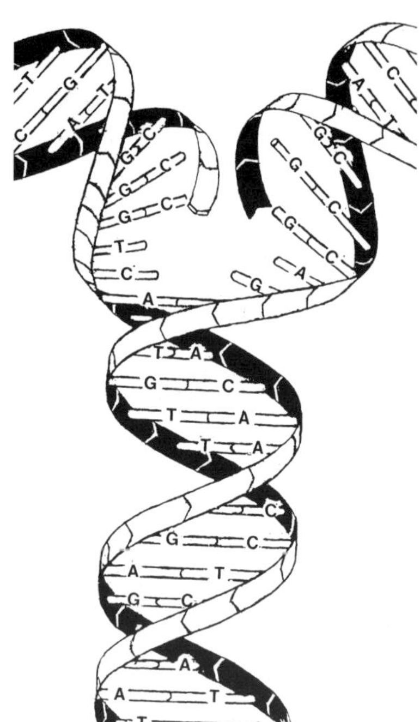

10 (*left*) 'Lucy', the skeleton of a female *Australopithecus afarensis* discovered by Donald Johanson in 1975.

11 (*above*) Diagram of the DNA molecule, shown reproducing itself. The A bases always link with T bases, and Gs with Cs.

12 (*above right*) The fruit fly *Drosophila melanogaster,* female (*left*) and male (*right*) (*see* Genes on Chromosomes; Homeobox Genes).

13 (*right*) DNA. James Watson photographed in 1994 with the model of the DNA molecule that he and Francis Crick had built in 1953.

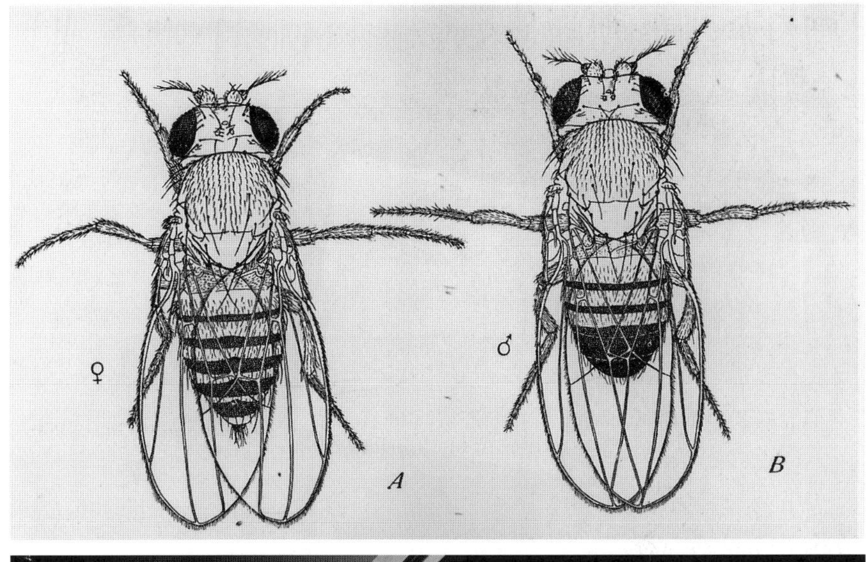
♀
A
♂
B

DOUBLE HELIX MODEL FOR DNA

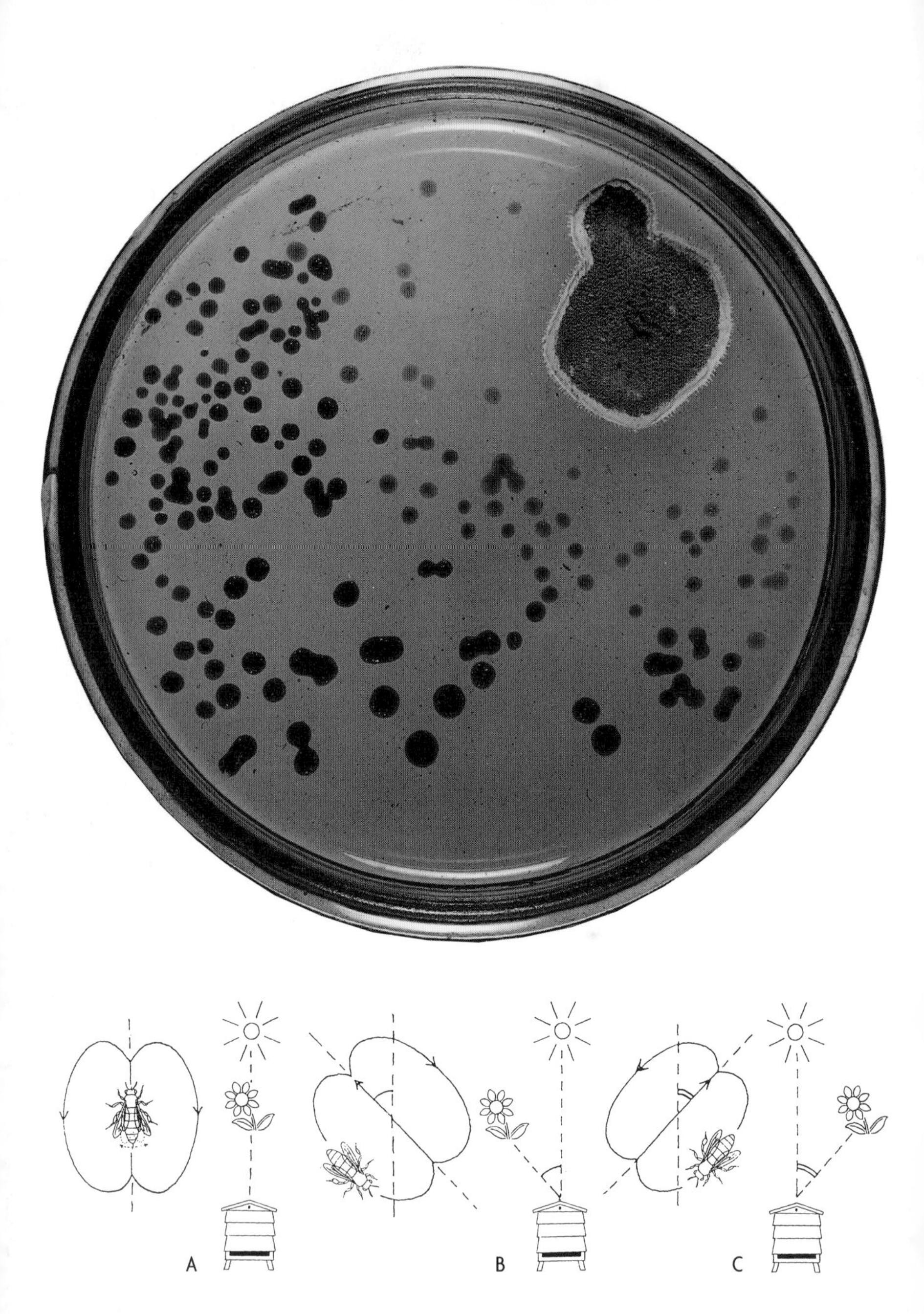

14 (*top*) The discovery of penicillin. The bacteria colonies in this dish (small round blobs) are being killed by a chemical coming from the mould *Penicillium* (large blob, *top right*).

15 (*above*) Bee dances. Performed on the vertical surface of a honeycomb, these show the direction in which flowers are to be found, relative to the direction of the Sun.

into actual physical contact with each other in order to do this. The simplest explanation was that live bacteria were occasionally coming together and exchanging their genomes bit by bit.

Further studies of this phenomenon showed that almost all of *E. coli's* genome is contained in one big loop of DNA, but there is sometimes another small loop, called a 'plasmid'. The plasmid contains genes that cause the bacterium to grow a little tube, with which it will inject the plasmid into another bacterium. This process is called 'conjugation'. Just occasionally, the plasmid gets spliced into the main loop of DNA, and when this happens, part or all of the main loop can get injected into another bacterium along with the plasmid itself. This is 'recombination'. It takes about ninety minutes for a bacterium to transfer its entire genome, but the process usually gets interrupted before then, so only part of the genome gets transferred. Experimenters were quick to see the possibilities this opened up. One can make a map of the locations of all the genes on *E. coli*'s genome, by seeing how many minutes two bacteria must conjugate for before each gene is transferred. (This is done by mixing two cultures with different identifiable genes, and then shaking them off each other with a kitchen blender after a set length of time.) One can also use this method to observe how the action of one gene is affected by the presence or absence of another. It was this discovery, as much as any other, that kick-started the revolution in genetics over the next fifty years.

J. Lederberg and E. L. Tatum, *Nature*, 158, 558 (1946)

TRANSPOSONS

Jumping genes

Science is never more romantic than when an outsider is proved right. Barbara McClintock was always a respected geneticist, working at some of the most prestigious laboratories, but for thirty years her research met with incomprehension and disinterest. Starting in the 1940s she worked tirelessly on the genetics of plants, especially maize, during the years when most geneticists were concentrating on bacteria and viruses. Most could not understand why she chose to study inheritance in organisms which only produce one new generation each year, when everyone else was working with flies whose life cycle repeated every ten days, or microbes that reproduced every twenty minutes. It just seemed so perverse of her. However, her project yielded a discovery with enormous implications.

This discovery was that not all genes remain forever located at the same place on their chromosomes. There are pieces of DNA that can hop around from one place to another, with great effect on other genes in their immediate neighbourhood. Some of these 'transposons' actually contain the code for particular proteins, but many of them chiefly affect other genes by switching them off. Since they can switch off several genes at once, their effects can be considerable. They are found in all living things, including bacteria. Indeed, one of the major discoveries of the Human Genome Project is that a major fraction of a human's DNA consists of transposons. However, McClintock had good reason to study them in maize. Ever since the 1930s she had been building up expertise in identifying particular regions on a maize plant's chromosomes—a very difficult achievement. With this knowledge, she could spot when a gene had moved. The particular genes she studied affected the colour of the kernels on a corn-cob. Sometimes a corn-cob can have multicoloured kernels, showing where a transposon is moving around, causing pigment genes to be switched on and off. Later, she also studied genes affecting the colours of leaves and flowers in other plants. She once explained the importance of genetic switches by saying: 'With the tools and the knowledge, I could turn a developing snail's egg into an elephant. . . . It is a matter of timing the action of the genes.'

It was only in the 1970s that other geneticists came to realise the full importance of transposons. The excitement started when they were found in bacteria, and it became apparent that they could move from one bacterium to another. This happens by means of 'bacterial sex', when bacteria exchange small loops of DNA called 'plasmids'. A transposon can attach itself to a plasmid, and then get transferred to another bacterium. When this happens, a whole set of characteristics could spread through a population of bacteria at great speed, much faster than an ordinary mutation. This was how bacteria could suddenly acquire resistance to a whole string of antibiotics—causing major problems for the treatment of diseases. Even more worrying, transposons could insert themselves into viruses, which would then carry them from one species of organism to another.

B. McClintock, *Cold Spring Harbor Symposium*, 16 (1951)

THE DNA DOUBLE HELIX

'The most important discovery since Darwin'

As soon as it was known that genes were made of DNA, the next question was: What is the structure of the DNA molecule? It must be the strangest molecule in existence—one that can carry in code the entire design specification for a living organism, and can run off copies of itself. How could a molecule do that? Among the researchers working on the problem were Maurice Wilkins and Rosalind Franklin at King's College, London, and Francis Crick and James Watson at Cambridge. In 1953, Crick and Watson were eager to beat everyone else to it (although both of them were meant to be working on other projects at the time). They were almost too impatient.

The data needed to find the structure of an organic molecule comes from X-ray diffraction analysis, but Crick and Watson were not experts at this technique. So they took a more theoretical approach. They knew that DNA was built of subunits: ribose, phosphate, and the 'bases' adenine, cytosine, guanine and thymine. So they constructed little aluminium models of these subunits, and started attaching them to each other, trying to find a way to put them together that fitted their knowledge of the molecule's unique properties.

They soon found they needed more data, in the form of better X-ray diffraction pictures to check their models against. Rosalind Franklin in London was producing these with meticulous care, convinced that there were no short cuts to solving the problem. She was not keen to be distracted by the model-builders at Cambridge. So what James Watson did next has raised many a scientific eyebrow. Without Franklin's permission, he persuaded her professor Maurice Wilkins to let him have look at her pictures, and then returned to Cambridge, where he and Crick quickly produced their famous 'double-helix' model (*see* Plate 12). They showed it triumphantly to Franklin, who immediately saw that it was indeed the correct structure to fit her data. Crick and Watson published their structure in a quick paper in *Nature*, telling everyone that it was 'the most important discovery in biology since Darwin's theory of evolution'. The same issue of *Nature* also contained papers by Franklin and Wilkins, describing the data.

Everyone was very excited when they saw the structure of DNA, because they could instantly see how it could perform its task: to carry coded information, and to produce copies of itself. The code comes in the form of the sequence of bases adenine, cytosine, guanine and thymine. The two strands of the double helix are only weakly joined to each other, and can easily pull apart. Each adenine base on one strand links to a thymine base on the other, and each cytosine links to a guanine. To produce a copy of a DNA molecule, the two strands of the double helix must be pulled apart. A new 'complementary' strand is then built along each separate strand with thymine bases linked to adenines, and cytosines linked to guanines, thus producing two identical DNA molecules where previously there was only one (*see* Plate 11).

J. Watson and F. Crick. *Nature*, 171, 737–738 (1953)

THE CENTRAL DOGMA

Genetic information flows from DNA to RNA to proteins

The 1950s was the most exciting decade for geneticists. Firm experimental results were coming slowly, but theories were evolving fast. And among the theorists, none was more enthusiastic than the extrovert Francis Crick, co-discoverer of the DNA double helix, and a man about as far from the boring-white-coat popular image of a scientist as could possibly be. He later described it as a time when they knew very little for certain, but had 'a boundless optimism that the basic concepts involved were rather simple and probably much the same in all living things'. It was in this spirit that in 1958 he delivered a lecture to the Society of Experimental Biology containing a number of inspired guesses that have since been proved to be almost entirely right.

Crick's central suggestions were that genes specify the composition of protein molecules, and that 'once information has passed into a protein, it cannot get out again'. Genes make copies of themselves, but proteins do not. Genes provide the templates on which protein molecules are built, but you cannot use a protein as a template for a gene. This he grandiosely called the 'Central Dogma'. He said that 'the direct evidence . . . is negligible', but it was the only way he could find to make sense of the data then available. Crick's main guesses went as follows.

- In a gene made of DNA, the genetic code is carried in the sequence of the four different kinds of base (usually called A, C, G and T) from which it is made. These code for the sequence of amino acids in a protein molecule.
- If the amino acids are strung together in the right sequence, the resulting protein molecule will automatically fold up into its correct final shape.
- The molecule ribonucleic acid, or RNA, which is found throughout living cells, is involved in transcribing the genetic code from DNA to proteins. RNA is a chain-like molecule similar to DNA, and it may be that an RNA chain is built using the DNA as a template, and the protein molecule is then built using the RNA as a template.
- There is probably more than one kind of RNA. In addition to the sort used for carrying information from DNA, there are other kinds involved in using this information to build protein molecules.
- This process probably goes on in tiny structures Crick called 'microsomal particles', which we now call ribosomes.

Over the past forty years, only two significant exceptions to Crick's scenario have been found. One is the enzyme Reverse Transcriptase, which allows genetic information contained in RNA to be transcribed back into DNA again; and the other is the 'prion', the protein involved in Mad Cow Disease, which can apparently make copies of itself without the need for any DNA or RNA at all. Crick had effectively deduced how living things work. His achievement is comparable to Charles Darwin's of a hundred years earlier, when he worked out the theory of evolution from similarly fragmentary evidence.

F. Crick. *Society of Experimental Biology Symposium*, 12, 138–161 (1958)

THE NATURE OF THE DNA CODE

Triplets of bases

In 1953, Francis Crick and James Watson had worked out the structure of DNA, the molecule that genes are made of. It is the famous 'double helix', a molecule of immense length, built like a spiral staircase, in which the individual steps consist of pairs of 'bases'. There are four types of base—adenine, cytosine, guanine and thymine, generally called A, C, G and T for short. Immediately they suggested that the genetic code was carried in the sequence of bases along one strand of the double helix. A single gene gives the specification for a single protein, and proteins are made of amino acids strung together. So it seemed a safe bet that the sequence of amino acids in a protein is specified by the sequence of bases in its gene. However, there are no less than twenty different kinds of amino acid found in proteins, so the next question was: How can the genetic code be written with just four different bases?

The answer was deduced by a team led by Crick, who studied the DNA of bacteriophages, the viruses that attack bacteria. These carry all their genes on one long DNA molecule. Crick induced mutations in a gene, using chemicals that he (rightly) believed would insert one or more extra bases into it. He found that adding one or two bases completely scrambled the code, so that the gene in question failed to specify a working protein molecule. But adding three bases simply resulted in a protein with a few 'wrong' amino acids in it.

Crick explained this result with one of his impressive chains of close reasoning. The simplest explanation, he said, goes as follows. The machinery that transcribes genes into proteins starts at one end of the gene, and reads off the bases in threes. Each 'triplet' of three bases specifies one amino acid. With four different bases available, there are sixty-four different possible triplets, such as ACG, ACT, GTC, GTA, which is more than enough to specify twenty different amino acids. A single DNA molecule can carry many genes, so there also has to be a triplet that specifies where one gene ends and the next begins. All sixty-four possible triplets are used, so each amino acid can be specified by up to three different triplets. If one artificially adds one or two bases to a gene, all the triplets from that point onwards specify the wrong amino acids. But if one adds three bases, one has effectively added just one extra triplet; only the part of the gene lying between the first and the last addition is scrambled, while the rest is left unchanged.

Over the next few years, the code was deciphered, and it is now possible to read a gene and know the composition of the protein it specifies. It is also possible to put together a wholly artificial 'designer' gene to specify any protein one wants. This has opened up the whole field of genetic engineering, with all the exciting and troubling possibilities it brings.

F. Crick et al, *Nature*, 192, 1227–1232 (1961)

THE OPERON

How genes are switched on and off

One of the first things to be discovered about genes was that there is a complete set of them in nearly every one of the millions of cells in our bodies. A cell in the skin on one's little finger, for instance, contains the complete recipe for growing an entire human being. And yet, clearly not all these genes are being used at the same time. They are being switched on and off, as appropriate. How this is done has been the major question in genetics during the last fifty years. The picture is immensely complicated, but some general features became clear quite early on.

Probably the biggest early breakthrough came in a blockbuster of a paper by two French researchers, François Jacob and Jacques Monod, working at the Institut Pasteur in Paris. They were working on bacteria, specifically the famous *Escherichia coli*. (*E. coli* has received bad publicity in recent years, because some mutant strains can cause fatal food poisoning. However, most strains are completely harmless, and huge numbers of them live peacefully inside the intestines of every one of us.) Their starting-point was to feed cultures of bacteria on the sugars glucose or lactose. The bacteria preferred the glucose, and only switched to feeding on lactose when the supply of glucose ran out. In order to digest lactose, they had to generate an enzyme called beta-galactosidase, which they would only do if there were lactose present but no glucose. It took them about twenty minutes to make the switch.

Jacob and Monod's discovery was that the gene that specifies the beta-galactosidase enzyme is positioned next door to two other genes, called the 'repressor' and the 'operator'. The repressor specifies a protein that sticks to the operator, jamming the machinery that transcribes the genes into enzymes. But if lactose is present, this repressor protein comes unstuck, and sticks to the lactose molecules instead. The machinery is now unjammed, and the enzyme's gene is now free to start generating enzyme. There is also a fourth gene which works in a broadly similar way, keeping the entire lactose system switched off as long as there is glucose present. The entire set of four genes is called an 'operon'.

The difficulty with which even this fairly simple system was teased out can be imagined. It entailed using the ability of bacteria to swap genes by means of 'bacterial sex', so that mutant versions of all four genes could be put together in different combinations, to see how they affected each other. Working out the mechanism of just one operon, however, is only a tiny beginning. The major question is: Do all genes in all living things work this way? It has been found that broadly speaking, they do. In bacteria, simple operons are the norm. In more complex organisms such as peas, fruit flies, or ourselves, the repressor genes are not always neatly positioned next door to the genes whose actions they regulate, but the principles involved are very similar.

F. Jacob and J. Monod, *Journal of Molecular Biology*, 3, 318–356 (1961)

REVERSE TRANSCRIPTASE

The enzyme for 'designer genes'

During the decade after Francis Crick and James Watson discovered the structure of the DNA molecule, the working of the genetic mechanism was slowly unravelled: how the genetic blueprint contained in an organism's DNA is transcribed to make the organism itself. The vital intermediary is a molecule called 'mRNA' (messenger ribonucleic acid), very similar to DNA, which carries the information in a single gene from the DNA to the microscopic machinery that reads it off and builds an enzyme molecule. Crick surmised that genetic information always moved along this one-way-street, from DNA to mRNA to enzyme.

Crick was largely right, but in 1970 a major exception to his rule was found. In the same week, the journal *Nature* received papers from two of the world's most noted biologists, David Baltimore at the Massachusetts Institute of Technology, and Howard Temin at the University of Wisconsin. Both had simultaneously, and quite independently, discovered that genetic information carried in an mRNA molecule can be translated back into DNA. They found a special enzyme, called Reverse Transcriptase, that does this. All living things have it, and use it as part of their machinery for correcting any copying errors that may appear in their genes.

The key to this discovery was a virus called the Rous Sarcoma virus (named after Peyton Rous, who discovered it in the early twentieth century) which causes cancer in chickens. It had long been known that this virus (like several others) had genes made of RNA, instead of the usual DNA. It was in this virus that Baltimore and Temin discovered reverse transcriptase. When the virus infects a cell, its genes are translated into DNA that is then incorporated in the cell's own chromosomes. It was soon found that many viruses transmit their genetic information in this way. They are called the 'retroviruses', and it is particularly difficult to cure or vaccinate against the diseases they cause. Once their genes have been transcribed into DNA and incorporated in a cell's genome, they are very hard to attack. The HIV virus that causes AIDS is probably the most notorious retrovirus. One of the early surprise findings of the Human Genome Project has been that a very high proportion of all our genes are not really our own, but are relics of old retroviruses, against which we have evolved a complete immunity.

The discovery of reverse transcriptase made possible a key technology in genetic engineering—identifying a gene that specifies for a particular protein, and then making copies of it to order. Normally, genes are hard to get at. Any given gene is hidden away among thousands of others, locked up inside a cell's nucleus. However, its transcript in RNA is much easier to isolate. When we read reports of scientists 'isolating the gene for' some particular characteristic, often they have in fact isolated its RNA transcript. Using reverse transcriptase, they can then use this transcript to construct a copy of the original gene in DNA.

D. Baltimore, *Nature*, 226, 1209–1211 (1970)
H. M. Temin and S. Mitsuzani, *Nature*, 226, 1211–1213 (1970)

ONCOGENES

Mutations that cause cancers

The causes of cancer are proving particularly hard to unravel. When a healthy tissue turns cancerous, its cells start proliferating uncontrollably. This change can be triggered by a number of agents, such as ultraviolet light, or the tars in tobacco smoke. So the question is: How do these agents cause a cancer to start growing? It seemed a reasonable guess that they must damage one or more of the genes which are involved in the control of tissue growth, but it was only in 1969 that such a gene was clearly identified.

The first of these 'oncogenes', as they are called, was found not in a cancer, but in a virus. The Rous Sarcoma virus has been known since the early years of the twentieth century, when it was found that certain cancers in chickens were spread by some infectious agent. In 1969 Steven Martin at the University of California, Berkeley, performed experiments to find out which genes in the virus had this cancer-causing effect. He subjected the virus to massive doses of the chemical nitrosoguanidine, which was known to cause genetic mutations. Most of the virus particles were killed off by this treatment, but just a few remained infective. Among these were a tiny number that appeared normal in every respect except that they failed to cause cancers. Apparently the nitrosoguanidine had caused a mutation in just one gene that gave the virus its cancer-causing ability.

When the gene was later analysed, it was found to be a truncated form of a gene belonging to the chickens that the virus infects. Virus particles reproduce themselves by injecting their genes into the tissues of their host (in this case, a chicken). The genes replicate there and re-program the host's tissues to make more virus particles. Presumably at some time in the distant past, part of one of the chicken's own genes had got mixed up in this process, so that copies of it had ended up in the virus particles. The gene in question is one that plays a part in regulating the growth of tissues. The shortened version carried in the virus can overwhelm the full-length version carried in chicken tissues, so tissue growth is no longer regulated at all, resulting in cancer.

Since this discovery was made, many more oncogenes have been found. To start with, they were found in viruses like the Rous Sarcoma virus, but later they were also found in cancerous tissues. Transplanting such a gene from a cancer cell to a healthy cell would cause it to become cancerous. There seems to be a number of genes which, when damaged by a cancer-forming agent or by being incorporated into a virus, become out of control, so that they cause tissues to grow endlessly, resulting in cancer. These genes are now being intensively studied, not just because they may hold the key to preventing or curing cancer, but also because they may tell us a great deal about how the growth of organs and tissues is normally regulated.

G. S. Martin, *Nature*, 227, 1021–1023 (1970)

SEQUENCING A GENOME

The complete genetic blueprint for a virus

To read the entire design specification of a living organism must be one of the most ambitious goals scientists have ever set themselves. In the first sixty years of the twentieth century, it was established that an organism's 'genes'—the specifications for all the different kinds of protein molecule that control its workings—consist of lengths of DNA, a chain-like double molecule built like a twisted ladder, whose rungs consist of pairs of four different types of 'bases', usually written as A, C, G, and T. The exact order of these four bases specifies the sequence of amino acids—the building blocks of the protein molecules. So an organism's genome, its entire set of genes, consists of an immense sequence of bases. Now the question was: How best can we read this sequence in the laboratory? From the early 1970s onwards, various methods have been used to read the sequence of individual genes. The first method that could read off bases fast enough to be practical for sequencing an entire genome was devised by Frederick Sanger and A. R. Coulson at the Medical Research Council Molecular Biology Laboratory, Cambridge, in 1975. Less than two years later, they were able to publish the complete genome of a virus, the first organism to have its design specification written out in full.

The virus in question was a 'bacteriophage'—one that attacks bacteria. Sanger and his team chose it because it is one of the simplest of all organisms, being made of only nine different proteins. They called their method the 'plus and minus' technique. The 'minus' part was to make four different copies of the genome, each one missing one of A, C, G, or T, and seeing where the breaks in each of these defective copies occur. The 'plus' part complemented it: an intact genome was broken down, using a system that creates a break wherever an A (or a C, or G, or T) occurs. These two procedures check each other. However, even the tiny genome of a bacteriophage would be far too big to analyse all at one go in this way, so it had to be broken down beforehand into chunks of manageable size. This was done by using enzymes called 'Restriction Endonucleases', which home in on particular short sequences of As, Cs, Gs and Ts, and cut the DNA molecule wherever they occur. There are several different kinds, each specific to a particular sequence. They are among the most powerful tools that geneticists have at their disposal, being used whenever it is necessary to isolate an identified length of DNA.

Sir Frederick Sanger is one of the very few people to have won two Nobel Prizes. (The first was for his earlier work on sequencing proteins.) His team's achievement was chiefly significant for the possibilities it opened up. The genome of a tiny bacteriophage might be of limited interest, but by sequencing it the team demonstrated a technique that could be used on any organism, including human beings.

F. Sanger et al, *Nature*, 265, 687–695 (1977)

THE ANTIBODY PROBLEM

How can we produce an antibody for every disease? By re-shuffling a pack of genes

It is common knowledge that whenever our bodies encounter a virus or a bacterium that causes disease, it will manufacture a specific 'antibody' for it—a protein that attaches itself to the outside of the germ, immobilising it. It is our bodies' first step in fighting back against an infection. But how do we produce the right antibody every time? Every different germ needs its own special antibody that will bind to it and to nothing else. We cannot possibly have a specific gene for every possible antibody—the variety of antibodies we can produce has to be very nearly infinite. This problem had been understood ever since antibodies were first described in the nineteenth century, but the explanation only came in 1983, when Susumu Tonegawa of the Massachusetts Institute of Technology described a model that was immediately recognised as correct. Some discoveries are the result of a single experiment. Tonegawa, by contrast, quoted evidence from no fewer than 119 different experiments performed by many different people (including himself), to piece together the full picture.

Antibody molecules are manufactured by our white blood cells, specifically by a particular type called 'lymphocytes'. Different individual lymphocytes produce different antibodies. Tonegawa's discovery was to show how they can do this with only a limited number of genes to provide the blueprints.

All antibody molecules, he said, have the same basic plan. However, the 'business end' of each molecule, which actually does the work of attaching the molecule to an invading germ, is made up of ten smaller parts, each with its own separate gene. Each of these ten genes comes in more than one version—there may be as many as three hundred different versions present in a single lymphocyte. Every time a new antibody molecule is produced, a different combination of these versions is used. This alone allows one lymphocyte to make an enormous variety of antibodies. But this is only the first part of the story. Tonegawa also found that the ten different genes are constantly mutating, so each lymphocyte comes equipped with its own unique set. This makes our lymphocytes able collectively to produce a limitless range of antibodies, with only a limited number of genes to work from. It is as if the genes come in packs, which are constantly being re-shuffled, and what is more, the cards in each pack are constantly changing.

Tonegawa's paper was one of the most exciting events in twentieth-century immunology. However, it does not leave the subject fully explained. We now know how antibodies are made. But we still do not understand how our bodies control the process. When we contract an infection, our lymphocytes produce more and more different kinds of antibody, until one is found that is effective in fighting it. This antibody is then made in large quantities, while other useless antibodies are discarded and forgotten about. We still do not understand fully how this is done.

S. Tonegawa, *Nature*, 302, 575–581 (1983)

HOMEOBOX GENES

Ancient genes controlling animals' body plans

An obvious feature of many animals' bodies is that they are divided into a number of roughly similar segments. They are most obvious in earthworms and centipedes, but a quick glance at other animals such as insects or lobsters is enough to show that they too have segmented bodies. Even vertebrates such as ourselves are segmented: they have repeating vertebrae running the length of their bodies, usually with similar ribs and blocks of muscle attached to each. (One can see them clearly in any fried fillet of fish.) Nobody believes that fish are descended from lobsters, or vice versa; but did they both inherit their segmented body-plan from a distant common ancestor? An unexpected discovery by geneticists in 1984 suggests that they probably did.

Walter Gehring is a Swiss geneticist who, like so many others, has used the fruit fly *Drosophila melanogaster* to study how an organism's genes are translated into its body-plan. In the early 1980s he assembled a team at the University of Basel to study some genes which can give rise to characteristic deformities: extra wings or legs, too many or not enough body segments, and so on. These so-called 'homeotic' genes clearly regulate just where in the fly's body other genes are switched on and off. The team soon made one intriguing discovery. The homeotic genes that control the design of the fly's body segments all occur together in a sequence on one of the fly's chromosomes. In the fly's head, only one of these genes is switched on, but as one progresses down the body more and more come into action, until they are all functioning at the tip of the tail.

This was interesting enough, but the really exciting discovery came next. In 1984 a new member of the team called William McGinnis found that many homeotic genes contained a small portion that was virtually identical. The team called it the 'homeobox'. The clear implication was that insects owe their segmented body-plan to a gene that has repeatedly been duplicated over the course of evolution, with each successive duplicate causing the growth of an extra segment. The team then started looking for the homeobox in other animals, and found that it is quite ubiquitous. Homeobox genes control how many vertebrae we have in our backbones, and also how many legs there are on a centipede. The eyes of a mouse and of a fly have completely different designs, but the same homeobox gene controls where (and how large) they grow on the animal's body. Even plants and fungi have genes with homeobox-like portions, and they always control the switching on and off of many other genes.

The obvious conclusion to draw is that the genes controlling organisms' body-plans are extremely ancient, dating back to a time before plants and animals had diverged into separate lineages. And the segmented pattern of flesh in a fillet of smoked haddock has the same ancestral pedigree as that in the tail of the lobster sitting next to it on the fishmonger's slab.

W. McGinnis et al, *Nature*, 308, 428–433 (1984)

DNA FINGERPRINTS

Easy-to-analyse differences between different peoples' genes

During the late 1980s, police and lawyers started using a highly exact technique for establishing a person's identity, commonly called 'DNA fingerprinting'. It was widely reported how a single hair found at the scene of a murder, or the smallest trace of semen found on a rape victim, could provide a sample of DNA from which it is possible to identify the criminal. This may not seem very surprising, since it is common knowledge that every individual human being has a unique set of genes, at least slightly different from everyone else's. So one might imagine that a technique is being used that provides a quick read-out of the genes present in a sample of DNA. In fact, this would be an impossibly huge task. The true method used is much simpler, and comes from a discovery that was made quite by accident.

In the early 1980s it was found that by no means all of our DNA actually contains genetic information. Interspersed throughout our genes are enormous lengths of DNA that do not code for proteins at all, and are (apparently) never read off by the machinery that translates the DNA specifications into actual tissues and organs. When this was first discovered, there was much talk of 'junk DNA', with the assumption that most of it is just there for the ride, because our bodies have never evolved a way of editing it out. More recently, geneticists have become more cautious about making such assumptions, and there is growing evidence that this supposedly useless DNA may play some role, possibly in controlling when our genes are switched on and off. It was this (supposedly) non-functional DNA that interested the team of Alec Jeffreys at Leicester University.

Jeffreys found that among the lengths of non-functional DNA were huge numbers of short pieces he called 'minisatellites', which could easily be identified because they consisted of highly repetitive sequences of the four 'letters' of the genetic code, A, C, G, and T. For instance, a typical minisatellite might contain long strings repeating the pattern 'GGAG GTGGGCAGGAGG'. Jeffreys devised a method of cutting all the minisatellites out of a person's DNA and analysing them on their own, and it was then that he made his unexpected discovery. In each person, the minisatellites come in a range of sizes, and this range is unique for each one of us. Using a standard laboratory technique called 'electrophoresis', Jeffreys could produce a pattern showing the different sizes of all the minisatellites in a person's genes. This pattern he called the person's 'genetic fingerprint'. He further found that each person's genetic fingerprint is about fifty-percent similar to the fingerprint of each of that person's parents. This means that they are useful in resolving paternity suits—one can easily see from two people's genetic fingerprints whether or not they are parent and offspring.

To this day, nobody knows what function (if any) minisatellites perform in nature. But Jeffreys saw their potential straight away. He immediately patented his technique, and a new industry was born.

A. J. Jeffreys et al, *Nature*, 316, 76–79 (1985)

'DOLLY', THE CLONED SHEEP

Cloning a mammal from adult tissues

Cloning had been a popular topic of speculation and unease for at least thirty years, since the possibility was first aired in popular books like *The Biological Time-Bomb*, written in 1968 by G. R. Taylor. So when a team of biologists led by Ian Wilmut at the Roslin Institute in Scotland announced in 1997 that they had cloned a sheep, it hit the headlines worldwide. The ethical implications were alarming. Few people were reassured by the fact that clones can occur naturally, in the form of identical twins. To most, running off copies of animals, or even human beings to order was deeply repugnant.

In fact, cloning was not entirely new. Cultivated plants, such as apple trees, have been cloned for thousands of years. Higher animals have been cloned ever since the early 1960s, when frogs were first cloned from cells taken from the intestines of tadpoles. The contents of the intestine cells were inserted into frogs' eggs whose nuclei (containing their genes) had been destroyed by high-energy radiation. This is relatively easy to achieve with frogs, because their eggs are quite large—about 3 mm in diameter. Achieving it with a mammal, whose egg cells are barely visible to the naked eye, would be much more difficult.

This is what Wilmut and his team had managed to do. They started with tissues taken from the mammary gland of a sheep, and teased out individual cells. From another sheep they obtained egg cells, and removed their nuclei. The mammary cells were then fused with the egg cells, by placing them in close contact and passing small electric currents through them. The resulting clone eggs were then grown in artificial cultures for a week before being placed in the uterus of a surrogate mother sheep. The end-result was 'Dolly', a lamb with no father, whose genes came entirely from tissues in a ewe's udder.

To biologists, the main point of this experiment was to see if all the genes carried in the various parts of an adult animal's body are still fully functional. A cell in a sheep's mammary gland contains all the sheep's genes, but clearly not all of them are being used. Are they all in full working order, and can they all be switched on again? Wilmut's experiment showed that the answer is yes. However, it was later found that the genetic material had been considerably damaged during the cloning process, which may explain why cloning usually does not work. Wilmut's team had to clone 277 eggs in order to get just 29 embryos. These they implanted into thirteen surrogate mother ewes, and Dolly was the only lamb to be born as a result. The experiment has been repeated many times since, but the success rate has scarcely improved. This fact alone makes it unlikely that humans will be cloned in the foreseeable future, even in countries which do not pass specific legislation to ban it. The armies of identical humans still belong in the realms of fiction.

I. Wilmut et al, *Nature* 385, 810–813 (1997)

ASTRONOMY AND COSMOLOGY

It can fairly be said that the twentieth century was the time when the cosmos entered into the public's consciousness, and caught its imagination. The rise of science fiction as a major genre of popular entertainment, the growth of amateur astronomy as a hobby, and the widespread (and highly unscientific) interest in UFOs all resulted from a shift in the way people viewed the world they lived in. Previously, people had thought of our planet as the Wide World, a big place: by contrast, in the twentieth century they started to think of it as Planet Earth, a small globe floating in an immense void. Two factors may be held responsible for this shift in perspective. The first was the arrival of modern rockets, making space-flight practical at least within the Solar System. The other, unquestionably, was the immense strides that were being made in astronomy.

This progress was driven very much by technical advances. The modern study of the Solar System began in the mid-nineteenth century, with the arrival of telescopes large enough to show the surfaces of the planets in some detail. By 1900, photography had progressed to a point where these telescopes could also be used to take lengthy time-exposure photographs of the stars, with the plates exposed for several hours at a time. By this means, distant stars and nebulae that had previously been too faint to see could now be studied in some detail.

And starting in the 1870s, astronomers also began to analyse the chemical composition of distant stars, using a device called a 'spectrograph', thus starting the science of astrophysics. A big turning-point came in 1925, when a telescope with a 250-centimetre diameter mirror, far bigger than any that had gone before, was installed at the Mount Wilson Observatory in California. This really marked the beginning of modern deep-space astronomy and cosmology. It was at Mount Wilson that Edwin Hubble first showed that the Milky Way is just one galaxy among many, and that the universe is expanding. Inevitably there arose the questions: How old is the universe? How did it begin? How will it end? Before Hubble, such questions could not even have been asked. Today, we have at least some answers.

After the Mount Wilson telescope, and its even bigger successors at Mount Palomar and elsewhere, the next major step forward was the beginning of radio astronomy in the late 1940s. (Radio waves were first found coming from outer space in 1933, but the Second World War delayed progress in studying them.) Previously, astronomers could only study hot, bright-shining objects such as stars and planets. The arrival of radio telescopes enabled them to look at cooler phenomena: the composition of interstellar gas and dust, and the faint 'cosmic background radiation' left over from the Big Bang. It also led to some surprises, such as the discovery of quasars and pulsars.

All these instruments are costly. Like physics, astronomy has become an immensely expensive science, relying almost entirely on state funding. Since it is unlikely to lead to any practical applications (unlike physics), one might wonder why governments are prepared to spend so much money on it. At least part of the answer really does seem to lie in simple awe at what astronomy is doing. Even complete non-scientists are instinctively fascinated by a science that seeks to ask: What is Out There? Where did it all come from? And where will it all end? There is a sense in which astronomy is seen to take over a job previously done by religion. Of all the sciences, it has the biggest 'wow! factor'. Not even politicians and civil servants are immune to it.

Of course, there is also one big question that everyone would like the astronomers to answer: Is there life out there? Will we ever meet intelligent aliens? As yet, the answer is that we simply have no idea. The possibility that primitive life may exist (or have existed) elsewhere in our Solar System is now considered strong enough to be worth investigating seriously. Mars and Europa (one of the moons of Jupiter) are currently believed to be the best places to look. But searching for life further afield is still barely practical. In the closing years of the twentieth century astronomers were starting to detect planets orbiting other stars, but looking for molecular oxygen in their atmospheres (the surest sign that life is present) is still beyond our abilities.

A search for intelligent aliens is even more problematic. It is a common belief, widely promoted by science fiction writers such as Arthur C. Clarke, that intelligent life must be widespread in the universe, and even in our own galaxy, and is just waiting for us to find it. This belief rests on three assumptions: firstly, that wherever the conditions are right, life will always arise spontaneously sooner or later; secondly, that wherever life arises, sooner or later intelligent beings will evolve; and thirdly, that wherever intelligent beings evolve, sooner or later they will show their presence, either by travelling in space or at least by making radio transmissions that we can detect. The truth is, we do not know whether any of these assumptions are justified, let alone all three of them. The evidence provided by the one example we know of, life on Earth, is quite inconclusive. So astronomers are probably justified in giving the search for intelligent aliens a rather low priority. It is probably something best done on the side, by sifting through data that was originally collected for other projects. Should aliens be found, it would be the most momentous discovery of all time, but among the ranks of professional astronomers no-one is holding their breath.

GALAXIES

'Island universes'

'Long, long ago, in a galaxy far, far away . . .' Today, every film-goer knows that the Milky Way, with its billions of stars, is only one of the countless number of galaxies that make up the universe. But as recently as the 1920s, this was news. Astronomers had catalogued many 'nebulae'—hazy patches of dim light that appeared in their telescopes, but they had no idea how far away they were, or even what they were. Best known of all was the Great Nebula of Andromeda, which is (just) visible to the naked eye (*see* Plate 6). By the start of the twentieth century, there were telescopes powerful enough to show that it was a clump of stars. But how far away was it? Was it a collection of dim stars, quite close to us, or of brighter stars, much further away?

The answer was provided in 1924 by a young American named Edwin Hubble. Born in 1889, Hubble had originally trained to be a lawyer. He studied first at Chicago University and then at Oxford University, where he acquired the style and manners of an upper-class Englishman. However, astronomy was his first love, and he decided to make it his profession soon after he returned to America in 1913. In 1920 he joined the Mount Wilson Observatory outside Los Angeles, which had just acquired a new telescope with a 250-centimetre aperture—then the biggest in the world.

Hubble was soon taking photographs of the Andromeda Nebula. In the winter of 1923 he saw something that excited him: a star in the Nebula whose brightness was changing from day to day. It was clearly a kind of star known as a 'cepheid', and that meant he could work out how far away it was. Cepheids are old, slightly unstable stars whose brightness fluctuates with clockwork regularity, and you can tell how bright a cepheid is by how fast it fluctuates. Big bright cepheids fluctuate more slowly than smaller, dimmer ones. So you can work out how far away a cepheid is, by comparing its apparent brightness as it appears in your telescope with its true brightness as you calculate it from its rate of fluctuation.

Hubble plotted the brightness of his cepheid from day to day, did the usual calculations, and came up with a figure: it was over a million light-years away. (A light-year is the distance light travels in a year, going at a speed of 300,000 kilometres every second). This made it easily the furthest object whose distance had ever been measured at that time. The Andromeda Nebula (and therefore, presumably, other similar nebulae that astronomers knew about) was far outside the Milky Way. The universe was clearly very much bigger than anyone had previously imagined.

Most scientific discoveries are first published in specialist scientific journals. Hubble's discovery did get published this way eventually, in 1925, in the *Astrophysical Journal*. But its first announcement was in the *New York Times* on November 23, 1924—not the first place one would normally look for cutting-edge science.

E. Hubble, *New York Times*, November 23, 1924; *Astrophysical Journal*, 62, 409 (1925)

THE EXPANDING UNIVERSE

Hubble's Law

During the first two decades of the twentieth century, several astronomers were studying the dim disc-shaped nebulae that had been found dotted around the sky. As more and more data was compiled, it was found that they all seemed to be moving either towards or away from us at very high speeds. This was discovered by the technique of spectroscopy—analysing the wavelength of light as if it were the pitch of a sound. Just as a racing-car's engine sounds high-pitched as it speeds towards us, but lower-pitched as it disappears off into the distance, so light's wavelength shifts towards the violet end of the spectrum if its source is moving towards us, and towards the red end if it is moving away.

Then in the 1920s, Edwin Hubble showed that the nebulae were distant galaxies, far outside the Milky Way. He measured (fairly accurately) the distances of a few of the nearest ones by observing some of the stars they contained, and found that the galaxies which appeared biggest and brightest were always the ones closest to us. The obvious explanation was that all galaxies are roughly the same size, and one can tell how far away a galaxy is simply from how small and dim it appears. Hubble then started to look at them with a spectroscope to see how fast they were moving towards or away from us, and soon made one of the most important discoveries ever in astronomy. *On aver-*

age, the further away a galaxy is, the faster it is seen to be moving away from us. This is now known as Hubble's Law.

At first glance it might seem that we must be at the exact centre of the universe, with everything rushing away from us. However, this is not the case; the universe would look much the same if seen from any place in it. Everywhere, the galaxies would seem to be spreading out from each other. Until recently it was thought that this expansion was slowing down under the influence of gravity, and that the entire universe might one day collapse back in on itself. However, the very latest observations suggest that this may not be true after all. Rather, the expansion actually seems to be speeding up, as if some as yet unknown force is causing the galaxies to repel each other.

Hubble's Law is notable because it confirms Einstein's theory of General Relativity. In the original version of the theory, the universe could not be static: it must either be expanding or contracting. Einstein was persuaded to revise the theory slightly to make it allow for a static universe, but when he learnt of Hubble's discovery, he said this was 'the biggest blunder of his life', and immediately went back to his original version. However, like all great discoveries, Hubble's Law poses bigger questions than it solves. If the universe is not static, does this mean it had a beginning? And will it have an end? These questions have dominated cosmology ever since.

E. Hubble, *Proceedings of the National Academy of Sciences, USA*, 15, 168–173 (1929)

PLUTO

The ninth planet

It is not often that the world's newspapers splash the latest astronomical discovery on their front pages, but that is what they did in 1930 when a ninth planet was announced.

Ever since Neptune was discovered in 1846, people have wondered whether there are any more planets waiting to be discovered. It was a particular fascination of Percival Lowell, a wealthy American who had founded an observatory at Flagstaff in Arizona with his own money. Calculating from data on the orbits of the two most distant planets known at the time, Uranus and Neptune, he claimed that they were being distorted by the gravitational attraction of a ninth planet orbiting somewhere out beyond them. We now know that Lowell was wrong, but after his death in 1916 the Head of his observatory, Vesto Slipher, took the idea very seriously. In 1929, he set the observatory's newest employee to work on a major search. The young novice was a farmer's son from Kansas called Clyde Tombaugh.

It is often told that Tombaugh pointed his telescope at the precise spot where Lowell had calculated that 'Planet X' should be, and lo and behold, there it was. This is quite untrue. Tombaugh spent nearly a year taking hundreds of photographs of the sky throughout the plane where the planets orbit. His method was to take pairs of photographs a few days apart, and then look at them with a 'blink microscope'. This instrument shows the two pictures in rapid alternation, so any star or planet that has moved is instantly noticeable. It is an immensely tedious task. Eventually, two photographs taken on January 23 and January 29, 1930, showed a 'star' that had moved. It was Pluto (*see* Plate 9).

The discovery was little more than a colossal fluke. To start with, Lowell's calculations were based on inaccurate data, and in fact neither Neptune's nor Uranus' orbits are distorted by a ninth planet. Besides, Pluto is tiny—its diameter is only about 2400 kilometres, which is about two-thirds that of the Moon, and it is very light even for its size. Its feeble gravity could not possibly affect the other planets. Finally, its orbit is not even in the same plane as the other planets, so it was pure good fortune that Tombaugh pointed his telescope at it at all. We now know that Pluto is the largest of the 'ice dwarfs'—small bodies of ice and rock that lie in the outer reaches of the Solar System. Most of the known ice dwarfs are satellites of the planets Jupiter, Saturn, Uranus and Neptune, but in recent years others have been found in orbits of their own, and there are surely many more waiting to be discovered. Pluto is in fact a double planet; it has a satellite called Charon, about a third its size, which was only discovered in 1978.

Pluto is the only planet in the Solar System that has never been visited by a space probe, although there are plans to send one in the coming decade.

C. Tombaugh, *Lowell Observatory Observation Circular*, March 13, 1930

WHAT MAKES STARS SHINE

Nuclear fusion reactions

'Twinkle, twinkle, little star, How I wonder what you are. . . .' People have always wondered what the Sun and the stars are made of, and what makes them shine. However, the subject remained a mystery until the first half of the twentieth century, when astronomers used the newly-invented technique of spectroscopy to find what stars are made of: mainly hydrogen, with some helium and much less of other elements. Meanwhile, the new science of nuclear physics at last suggested a powerful enough energy source: hydrogen fusion reactions, taking place under the intense pressures found in a star's interior regions. Stars are in effect gigantic H-bombs exploding in extreme slow motion.

The details were first worked out by the German-American Hans Bethe at Cornell University in 1938. According to his calculations, there were two processes involved, both resulting in four hydrogen atoms fusing to form one helium atom. In one process, two hydrogen nuclei, consisting of one proton each, fuse to form a deuterium nucleus, consisting of a proton and a neutron. This requires one of the protons to turn into a neutron, which results in energy being given off in the form of a fast-moving electron and an extremely light particle called a neutrino. Two deuterium nuclei then fuse to form a helium nucleus, consisting of two protons and two neutrons. In larger stars, however, another process predominates. These stars contain quantities of carbon, nitrogen and oxygen, which act as intermediaries. A continuous cycle occurs, starting with hydrogen and carbon, and ending with carbon and helium.

Bethe's calculations have since been confirmed, except that he was wrong about one major thing. He was under the impression that the universe was only some two billion years old. According to his calculations, a star could shine by turning hydrogen into helium for about twelve billion years before it started to run out of hydrogen, so he presumed that all the stars in the universe must still be quite young. Under these conditions he calculated that they could only generate helium, so all the heavier chemical elements such as carbon, oxygen, and the metals must presumably have been already in existence when the first stars formed. We now know that this is untrue. The universe is in fact over twelve billion years old, and originally it consisted of nearly three-quarters hydrogen and one-quarter helium, with tiny quantities of lithium also present. All the heavier elements are generated in old stars when their hydrogen supply starts to run out. The heaviest metals are generated in 'supernovae'—large stars that end their lives with a gigantic explosion as their internal regions collapse under their own weight.

Hans Bethe is one of the twentieth century's most versatile physicists, who has made major contributions in several fields. During the Second World War he was one of the chief architects of the atomic bomb, but he was never happy about nuclear weapons, and ever since he has been a highly vocal critic of American military policy.

H. A. Bethe and C. L. Critchfield, *Physical Review*, 54, 248–254 (1938); H. A. Bethe, *Physical Review*, 55, 434–456 (1939)

THE AGE OF THE SOLAR SYSTEM

Four and a half billion years

In the late seventeenth century, Archbishop Ussher calculated that, according to the Bible, the Earth came into being in the year 4004 BC. However, geologists were soon forced to the conclusion that it was considerably older than that. Its true age was finally worked out in 1955.

In theory, the way to find the age of the Earth would be to find a sample of rock that had survived undisturbed from the beginning of its history. However, due to the movements of the Earth's crust (see 'Plate Tectonics'), there are no rocks that old still accessible from ground level. So instead, scientists asked: How old is the Solar System? Presumably the Earth is the same age as the Sun and the other planets. Can we find material dating back to when they first formed? Indeed, we can—in the form of meteorites. These are fragments of asteroids, lumps of rock and metal that have been orbiting the Sun relatively undisturbed ever since the Solar System first condensed out of a cloud of dust. Occasionally they fall to Earth. One can estimate their age by analysing any radioactive isotopes they contain, and their decay products. For instance, uranium-238 decays to lead-206, and thorium-232 decays to lead-208. (Different isotopes of an element differ in the number of neutrons each atom contains.) The rate at which each of these isotopes decay is known with some accuracy. So by analysing the ratio of, say uranium-

238 to lead-206, one can work out how long the uranium has been decaying for—but only if one knows how much lead-206 was there to start with.

This was the problem solved by Claire Patterson of the California Institute of Technology. Meteorites come in two main types—stone and iron. Stone meteorites contain radioactive isotopes, but iron ones do not. However, iron meteorites do contain small traces of lead isotopes, which must have come straight from the primordial dust cloud. Patterson used the ratios of the different isotopes in this primordial lead to estimate the quantities that must have been present in the stone meteorites when they first formed. From this he was able to calculate how much of the lead in the stone meteorites was originally uranium or thorium, and hence how old they were. The age he calculated was 4.55 billion years. All meteorites that have since been studied have been found to be of this age, as are the oldest rocks brought back from the Moon by the Apollo astronauts (*see* Plate 7).

At four and a half billion years, the Earth is considerably older than people had previously suspected. One of the major arguments put forward by religious fundamentalists and others opposed to evolution theory was that the Earth was not old enough for evolution to have produced living things as varied and complex as we see today. When the Earth's true age became known, this argument lost much of its force, especially as evidence has since been found to show that life first appeared quite soon after the Earth first formed.

C. Patterson, *Geochimica et Geophysica Acta*, 7, 151–153 (1955)

THE ORIGIN OF THE CHEMICAL ELEMENTS

Inside old stars

When studying the history of the universe, one is immediately faced with the question 'Where did the chemical elements come from?' There are ninety-one different elements found in nature. Astronomers have always assumed that the lightest of these, hydrogen, is the primordial form of matter. Its atoms consist of just one proton and one electron (and occasionally one or two neutrons). In the 1930s it was calculated that the second lightest element, helium, was synthesised out of hydrogen inside stars. But how did all the other elements originate? Most scientists concluded that they must somehow have existed since the universe began, which they believed happened about two billion years ago.

However, a few astronomers disagreed. They did not believe that the universe had ever had a beginning. Rather, they propounded the 'Steady State' theory, which held that the universe had always existed, and that hydrogen is constantly coming into existence out of nothing, at a rate too low to detect. In 1957 the American astrophysicist William Fowler and the Englishman Fred Hoyle showed how this hydrogen is then converted into all the heavier elements, in a giant paper written in collaboration with Margaret and Geoffrey Burbidge.

These theorists realised that certain stars known as 'red giants' are in fact normal stars that have turned most of their hydrogen into helium and are nearing the ends of their lives. They showed that as the hydrogen-helium reactions wind down, the core region of a star will contract under its own weight, heating up as it does so. As the temperature rises, a whole series of further reactions occur, with atoms fusing to form successively heavier elements up to iron. Elements heavier than iron are then built up by further addition of protons and neutrons. All these elements are then spread around outer space as the star becomes unstable and starts blowing off its contents. Finally, the largest stars explode spectacularly as 'supernovae', creating the heaviest elements such as uranium in the process.

The strange thing is that Fowler's and Hoyle's explanation is actually the right one, even though their starting-point, the Steady State theory, was completely wrong. We now know that the universe did have a beginning, although this was some twelve billion years ago, rather than the mere two billion that had previously been believed. Originally it consisted of nearly three-quarters hydrogen, with about one-quarter helium and a very small amount of lithium. All the other elements originated inside red giant stars and supernovae, just as Fowler and Hoyle calculated.

Fowler was awarded a Nobel Prize for this discovery, but strangely Hoyle was not. Sir Fred Hoyle, who died while this book was being written, was a perverse genius, and other scientists' opinions of him varied considerably. He never relinquished his belief in the Steady State theory even in the face of overwhelming evidence against it. He also insisted that outer space is full of bacteria and viruses, which constantly rain down on the Earth—a notion considered madcap by biologists and astronomers alike.

E. M. Burbidge et al, *Reviews of Modern Physics*, 29, 547–650 (1957).

QUASARS

Giant black holes in deepest space

When astronomers first started using radio telescopes, they soon found small sources of intense radio waves, which they called 'radio stars'. It was already known that ordinary stars, such as our Sun, get their energy from hydrogen fusion reactions, like a giant H-bomb in slow motion. But no-one could think of a power source that could cause a star to shine with radio waves rather than with visible light. There was also uncertainty about how large and how distant these radio stars were. The early radio telescopes could only give the approximate direction of a radio source; they could not resolve a sharp image of it. Then in 1963, two teams of astronomers, working at the Parkes radio observatory in Australia and at the Mount Palomar telescope in California, joined forces to study two radio stars. Their results, published in four papers in the same issue of *Nature* magazine, were a complete shock.

The two radio stars had their positions determined with as much accuracy as possible, and the giant Mount Palomar telescope was then pointed at them. At first glance, they appeared to be small, dim stars. But then their 'red-shifts' were measured, to see how fast they were moving. It was found that they were moving away from us at immense speeds. According to Hubble's Law, the most likely explanation was that they were not nearby stars at all, but extremely distant objects, further away than the furthest galaxies known at that time. In order to be visible at all at such distances, they must be tremendously bright—brighter than entire galaxies, in fact. And yet, evidence soon emerged that they were very small—not much bigger than the biggest stars. Whatever radio stars were, they were clearly not stars at all, so they needed a new name. Their discoverers called them 'quasi-stellar radio sources', soon shortened to 'quasars'. It seemed that they were a feature of the distant past: when astronomers look at very distant objects, they are seeing light that started on its journey towards us a long time ago. There are no quasars near us, which means that today they no longer exist.

It was not until the early 1980s that astronomers came to agree on what quasars really were. It is now thought that they were almost certainly huge 'black holes'—objects so dense that nothing can escape their gravity, not even light itself. Their existence was predicted by General Relativity theory, but previously no-one was sure what they would look like. It is now believed that every galaxy contains a giant black hole at its centre. When the galaxies were very young, gas and even entire stars were constantly falling into these black holes, emitting vast quantities of radiation as they were accelerated in the intense gravitational field. In more recent times, the black holes have generally swallowed most of the matter in their immediate neighbourhood, so they are giving off very little radiation. Their presence can only be inferred from the movements of stars orbiting around them.

C. Hazard et al, *Nature*, 197, 1037–1039; M. Schmidt, *Nature*, 197, 1040; J. Oke, *Nature*, 197, 1040–1041; J. L. Greenstein and T. A. Matthews, *Nature*, 197, 1041–1042 (1963)

PROOF OF THE BIG BANG

Cosmic background radiation found by accident

In the decades that followed Hubble's discovery that the distant galaxies were speeding away from us, two rival theories emerged to explain the fact. The first was the 'Expanding Universe' theory, popularly known as the 'Big Bang'. This held that the entire universe had originally been unimagineably small, dense and hot, and it had been expanding and cooling ever since. Many people found this unsatisfactory, for aesthetic and philosophical reasons as much as anything else. One automatically asks: What was there before the Big Bang? What set it off? Where will this expansion all end? and several other questions.

So in the 1950s a few astronomers, notably Fred Hoyle, devised a rival 'Steady State' theory. According to this scenario, the universe is infinitely large, infinitely old, and has always looked the same as it does now. The galaxies are moving away because they repel each other. The universe does not empty out because new matter is very slowly coming into existence spontaneously throughout space, at a rate too low to detect. It is continually condensing and clumping together to form new galaxies. Many people found this theory much more satisfying.

The two theories made different predictions. The Expanding Universe theory said that the universe had changed over time, while the Steady State theory said that it had not. The way to test the theories was to look back into the past, which one can do by looking at objects very far away. The light reaching us from objects in the furthest depths of space set out on its journey towards us billions of years ago. According to the Expanding Universe theory, if one looks far enough, one eventually sees back to the time before matter was clumped into galaxies, and the whole universe was a seething soup of atoms. This should be seen as a wall of ancient radiation, now cooled down to form a faint wash of microwaves. Beyond that, everything should go opaque.

When this microwave background was actually found in 1964, it was quite by accident. Arno Penzias and Robert Wilson were two radio engineers working for the Bell Telephone Corporation, which was experimenting with microwave transmission for carrying long-distance telephone calls. They were using a giant horn-shaped antenna to pick up microwaves bounced off high-altitude balloons, in preparation for using satellites. However, they were also keen astronomers, and the Bell Corporation allowed them to use the antenna for radio astronomy as well. They soon found that it was picking up a uniform faint 'hiss' that seemed to come evenly from the whole sky. Assuming something was wrong with the apparatus, they checked it out meticulously, even eliminating the possibility that the radiation was coming from warm pigeon-droppings inside the antenna! Eventually they concluded that the hiss was real, and must be coming from deepest space. Astronomers soon realised it was what they had been looking for: the radiation left over from the time when the first atoms formed out of pure energy. It was proof of the Big Bang.

A. Penzias and R. Wilson, *Astrophysical Journal*, 142, 419–425 (1965)

PULSARS

The most densely compressed matter possible

Imagine a star rather larger than our Sun, which collapses under its own weight until its entire bulk is compressed into a sphere no more than nineteen kilometres across. One teaspoonful of it weighs three billion tonnes. Imagine that it is spinning on its axis at a rate of thirty or more revolutions per minute. Such an object is a pulsar.

The first pulsar was found quite by accident in the summer of 1967 by Jocelyn Bell, a member of a team at Cambridge University led by Anthony Hewish, which was using a new radio telescope to search the sky for small radio sources. One day she found a source that was emitting radio wave in discrete pulses, at a rate of exactly one pulse every 1.337 seconds. The team's first guess was that it was an artificial object, such as a space probe. However, further observations showed that the pulses were coming from far outside the Solar System. At this point, a sudden thought occurred: was this the first sign of extraterrestrial aliens? This possibility was soon ruled out, but for a while the team jokingly referred to the source as 'LGM', standing for 'Little Green Men'.

It took astronomers only a short time to work out that the source must be an object spinning at great speed, and sending out radio waves in a beam like the light from a lighthouse. The radio waves were so intense that only an object as massive as a star could contain enough energy to produce them, but this object was spinning so fast that it could only be a few kilometres in diameter. If it were any bigger it would fly apart. That meant it could only be one thing: a 'neutron star'. Neutron stars had been predicted by theory several years earlier. They are the remnants of 'supernovae'—old stars which collapse under their own weight. A supernova appears as a gigantic explosion, which for a few weeks can outshine all the other stars in a galaxy put together. But this colossal blast is just the outer layers of the star being blown off. Most of the star's mass falls inwards, compressing itself so tightly that the individual atoms collapse, with the electrons and the protons coalescing to form neutrons. The whole star has a strong magnetic field, and is also surrounded by an atmosphere containing many free electrons. The magnetic field causes the electrons to move in tight spirals, generating radio waves in two narrow beams, pointing in opposite directions and sweeping round as the star rotates.

Many hundreds of pulsars have now been found. It has recently been suggested that they may be one source of our heaviest metals, such as gold and platinum. The abundance of these metals is hard to explain, and it could be that they are formed in the explosion that occurs when two pulsars collide. These collisions must be extremely rare, but they would result in the most intense thermonuclear reactions.

A Hewish et al, *Nature*, 217, 709–713 (1968)

ORGANIC MOLECULES IN INTERSTELLAR SPACE

Possibility of extraterrestrial life

The one question which people would surely most like astronomers to answer is: Are we alone in the universe? Or is life widespread on other worlds? By the year 2000, astronomers still could not answer this question, but the chances of life arising elsewhere seemed to be high. The basic chemistry that leads to living things does not only occur on the surface of planets, but is actually widespread in outer space.

Life as we know it is built of 'organic' molecules—molecules with a central structure of carbon atoms, to which other elements are attached. By the 1950s it was already known that carbon and other elements are widespread in interstellar space. So the next question was: Does interstellar carbon form compounds? In the early 1960s, clear signs of simple inorganic molecules such as water and ammonia were detected in outer space, so could more complex compounds exist out there as well?

All chemical compounds could show their presence by absorbing radio waves. Each compound has its own unique 'footprint' of frequencies that it absorbs. So one way to look for compounds in interstellar space is to look at a distant radio source such as a quasar, and see what, if any, radio frequencies are failing to reach us. This was what Lewis Snyder and his colleagues did at the Green Bank Radio Observatory in West

Virginia in 1969. They looked at the radio waves coming from a number of known radio sources in deep space, and found that waves with a frequency of 4830 megaherz were being blocked out. The simplest chemical known to absorb radio waves at this frequency is the organic compound formaldehyde. However, radio waves coming from closer sources, such as nearby stars and the planet Jupiter, did not lack 4830 megaherz waves. Clearly the formaldehyde must be more distant. It appeared to be widespread in the thin gas clouds that are found throughout the Galaxy.

Formaldehyde is one of the simplest of organic compounds, each molecule consisting of just one carbon atom with two hydrogen atoms and an oxygen atom attached, but where there are simple organic molecules there could be more complex ones as well. In particular, formaldehyde can react with water and ammonia (both found in outer space) to form amino acids, the molecules that form the basic building-blocks of life. The hunt for more complex molecules was on, and over the next several years many were found. Finally in 1996 the amino acid glycine was detected. Very few astronomers believe that life could arise anywhere except on the surface of planets, but it is clear that when planets form out of the dust and gas surrounding young stars, several of the necessary raw materials for life are already present. So simple life may be quite widespread throughout the Galaxy. However, whether simple life often gives rise to more complex organisms, or even intelligent beings, is another matter altogether.

L. E. Snyder et al, *Physical Review Letters*, 22, 679–681 (1969)

GAMMA-RAY BURSTERS

The biggest explosions of all

In 1963, with the Cold War at its height, the American military were worried that the Soviet Union might secretly be conducting nuclear tests in outer space, perhaps even on the far side of the Moon. To check on this suspicion, they launched the 'Vela' satellite, designed to detect nuclear explosions in space by looking for the distinctive flashes of high-energy gamma-rays that they would produce. The satellite did not find any clandestine tests, but instead it discovered one of the most awesome phenomena in the whole of nature—gamma-ray bursters. Unfortunately, the satellite was top secret, and it was not until 1973 that the Pentagon felt confident enough to declassify the data, and allow civilian scientists to see it.

On average about twice a day, the satellite detected a strong flash of gamma-rays lasting for a few seconds, indicating some kind of nuclear explosion occurring somewhere in outer space. Clearly they could not all be Russian bombs, so they must be some natural phenomenon. In other words, they must be produced by very powerful explosions indeed, occurring very far away. But just how distant were they? To start with, most astronomers assumed that they must be coming from sources within our own Galaxy, simply because they were so intense. If they were more distant than that, the explosions would have to be quite incredibly violent. No physics known at the time could explain such a massive release of energy. But nonetheless this has since proved to be the case.

The main difficulty with studying gamma-ray bursters is their unpredictability. They occur completely randomly, and only last a few seconds. So it was not until May 8, 1997 that astronomers finally managed to train a telescope on one before it faded away. On that day, a specially-designed satellite called Beppo-SAX detected a gamma-ray burster and immediately relayed its exact location to the Mount Palomar Observatory in California. There, telescopes were trained on the spot and detected a rapidly-fading glow of visible light coming from it. Analysis of this light showed that it had taken ten billion years to reach us, making its source the most distant object ever detected. For the explosion to be visible at all at such a huge distance, it must have been far more powerful than anything previously envisaged.

Astronomers now believe that gamma-ray bursters must be the death throes of giant stars, perhaps ten or twenty times heavier than our Sun. Such stars may have been common ten billion years ago, when the universe was young. A giant star will burn out after only one or two million years, after which its interior will suddenly collapse to form a 'black hole', a point in space where the gravity is so intense that even light itself cannot escape it. This collapse triggers a colossal explosion, most of whose energy is channelled in two narrow beams of intensely powerful gamma-rays, pointing in opposite directions. We see a gamma-ray burster if one of these beams happens to point towards the Earth.

S. G. Djorgovski et al, *Nature*, 387, 876–878; M. R. Metzger et al, *Nature*, 387, 878–880 (1997)

PLANETS ORBITING OTHER STARS

Strange new worlds

It is the classic scenario in science fiction—our Sun is not the only star to have planets orbiting round it. There are so many stars in the Milky Way Galaxy that there must be lots more planets out there too, and a good many of them must be quite like our Earth. Most people have heard this so many times that they assume it is a known fact, but actually it is not. There is always the possibility that planets like the Earth are really very unusual. However, it was only in the closing years of the twentieth century that astronomers acquired instruments powerful enough to go looking for planets orbiting other stars. Their efforts were soon rewarded, but with some very surprising finds.

The problem with searching for planets orbiting other stars is simply that they are far too small to see, even with the very biggest telescopes. So astronomers have had to approach the problem indirectly. The main method has been to look for stars that wobble. If a star has a large planet (at least as big as Jupiter, say), its gravity will drag the star back and forth as the planet orbits around it. It should be just about possible to detect this, at least in some of the nearest stars. Everyone expected that the search would meet with success, but when it came, everyone was amazed. The planets were out there, all right—but not where they should be.

In 1992, two astronomers called A. Wolszczan and D. A. Frail were using the giant radio telescope at Arecibo in Puerto Rico to look for pulsars. These are the remnants of supernovae—stars that have exploded with unimaginable force. Any planets they had orbiting them should have been destroyed in the blast, or at least hurled out of their orbits. And yet Wolszczam and Frail found a pulsar that clearly had at least two planets orbiting round it. How they came to be there remains a complete mystery. This discovery encouraged astronomers to redouble their efforts to find planets orbiting more 'normal' stars, and in 1995 success came to Michel Mayor and Didier Queroz at the Geneva Observatory in Switzerland. But again, the planet was a big surprise.

All the theories about how planets are formed predict that small planets should be found close to their stars, while the larger planets should orbit further out, just as they do in our own Solar System. But the planet found by Mayor and Queroz was a giant at least as big as Jupiter, orbiting very close indeed to its parent star—so close that its atmosphere may be slowly boiling away. Again, no-one has a clue as to how it came to be there, but we already know that it is not unique. Several other giant planets have since been found in similarly small orbits.

So we are still left wondering: Are these newly-discovered planets very abnormal specimens—or are we?

M. Mayor and D. Queroz, *Nature*, 378, 355–358 (1997)

THE AGE OF THE UNIVERSE

Much younger than one might expect

One of the most impressive scientific achievements of the twentieth century has been the development of cosmology, the study of the universe taken as a whole. In 1900, questions such as 'How big is the universe?' or 'How old is it?' could hardly even be asked. A hundred years later, cosmologists were providing firm answers.

The evidence for the age of the universe comes from several different lines of research, which came together in the 1990s. In the last two decades of the century, several astronomical satellites were launched into orbit, including the giant Hubble Space Telescope, enabling astronomers to see the dimmest and most distant objects with far greater clarity than ever before. These instruments were used to answer the question: How far away are the distant galaxies? They are all rushing away from us and from each other as the universe expands, and we can measure their speeds quite accurately. If we could tell how far away they are, we could couple that figure with the speed at which they were moving apart, to calculate how fast the universe is expanding, and thus how long ago it was all massed together in a single point at the beginning of time. Several teams of astronomers used broadly similar methods to attack this problem. The biggest study was by a team headed by Wendy Freedman at the Carnegie Institute in Washington, who recorded exploding stars of a particular type called 'Type 1a Supernovae'. These explosions are extremely bright, so they can be seen even in very distant galaxies. It is also known that they are all of very nearly the same brightness. So one can accurately measure a Type 1a's distance from how bright it appears.

At first, the different studies gave conflicting results. Some observations suggested that the universe might only be eight billion years old, which was strange, seeing that the oldest known stars were believed to be up to twelve billion years old. But finally the differences were ironed out, and in 1999 Freedman announced that they were in broad agreement. (Her full report appeared in print the following year.) The entire universe is about twelve to fourteen billion years old. By comparison, our own Solar System is known to be about 4.5 billion years old.

This is a startling figure. When one considers the vast size of the universe, and how insignificantly small our own Solar System is in comparison, one might guess that the universe is immensely much older than the Earth. But it is not—it is only about three times as old. And for the first few billion years of its history, it consisted of about three-quarters hydrogen, one-quarter helium and a very small amount of lithium. The other chemical elements, such as carbon, oxygen, and the various metals, only appeared more slowly. So the Earth, with all its heavy elements, may belong to only the first or second generation of solid planets to have come into existence.

W. Freedman et al, *Physics Reports*, 333–334, 13–31 (2000)

EARTH SCIENCES

In the first decade of the twentieth century there appeared a few scientists who can now be seen as the originators of modern 'earth science': men whose research covered geology, meteorology, and anything else there was to be studied about the physical workings of planet Earth. Two of them stand out: Andrija Mohorovicic in Croatia, and Alfred Wegener in Germany. Mohorovicic is remembered for discovering the boundary between the Earth's outer crust and its underlying mantle. Wegener, meanwhile, in 1912 propounded a hypothesis that almost everyone dismissed as completely crackpot: that the Earth's continents are not static, but are slowly on the move. He was not the first to have noticed that the eastern coastline of South America and the western coastline of Africa appear to fit each other like two pieces of a giant jigsaw, but he was the first to suggest that that was precisely what they were.

With hindsight, it is easy to see that Wegener was right. However, when he first proposed his theory, he had little hard evidence to go on. That came later, as geologists slowly charted similarities between the rock formations on either side of the Atlantic Ocean, and then mapped the floor of the Atlantic itself and showed that it was getting wider at the rate of a few centimetres each year. Then in 1960, Harry Hess at Princeton University hit on the mechanism that was making the continents move: convection currents inside the Earth's apparently-solid mantle, which were causing the thin outer crust to move like the scum on a pan full of boiling jam, in extreme slow motion. (These currents had originally been suggested by the Dutch geophysicist Felix Vening-Meinesz nearly thirty years earlier, but he was before his time.) At last, Wegener's hypothesis could be called 'scientific'—it was complete enough that it could be used to make predictions that could be tested. The stage was set for the development of the Plate Tectonics theory, which finally took shape in late 1967 and 1968. Alone among the discoveries described in this book, it cannot be said to be the work of any one scientist or even team of scientists. Rather, it was arrived at by a consensus among the big majority of the geologists of that time. In this way, it is an extreme example of the collaborative nature of late twentieth-century science, a world away from the lone researchers of earlier times.

As in the other sciences, the immense progress in the earth sciences in the twentieth century resulted from the greatly increased resources poured into them, both by industry and by governments. Industry was chiefly interested in geology because of its applications to mining and especially prospecting for oil. After the First World War everyone could see that the world demand for oil was going to increase enormously, and the race was on to find supplies to meet this demand. Geologists proved their worth in the 1930s by predicting the existence of massive oil reserves in the Persian Gulf region, which were finally opened up in the late 1940s and 1950s. Meanwhile, governments were interested in earth sciences for a variety of reasons. They had been involved in weather forecasting ever since the 1850s; the arrival of radio made it possible for data to be relayed to meteorologists from remote unmanned weather stations, from aircraft, high-altitude balloons, and finally satellites in space. The discovery that the upper atmosphere could reflect radio waves, making it possible for radio signals to be sent around the globe, gave governments a further interest in atmospheric physics. It may at first sight seem surprising that much government-sponsored research in the earth sciences had military objectives. For instance, study of the Earth's gravitational and magnetic fields, as well as its upper atmosphere, was of great importance to the development of intercontinental ballistic missiles. This knowledge was needed in order for them to hit their targets accurately to within a few metres after being lobbed half-way round the world.

It was for reasons such as these that in 1956, sixty-seven nations around the world, including the USA and the USSR and their allies, jointly announced the International Geophysical Year (which actually ran for eighteen months), in which they collaborated in several different fields of study in the earth sciences. The nations signed a treaty renouncing all territorial claims in Antarctica, so that scientists could roam over it as they wished without being restricted by national boundaries. The Soviet Union amazed the world by launching Sputnik 1, the first-ever satellite, which provided data about the outermost regions of the atmosphere. It may seem strange that the biggest effort in international collaboration that scientists have ever achieved should have taken place at the height of the Cold War, but in fact military considerations were at the heart of it. The only way that the military establishments of both sides could obtain the data they wanted was to enlist the help of as many other nations as possible.

Earth scientists made immense progress during the last century, but on many practical matters they found themselves coming up against a limit. The field of mathematics called chaos theory originated partly in attempts to model global climate and weather patterns. It tells us that it will never be possible accurately to forecast the weather more than four or five days in advance, simply because the system is so complex. A more broad-brush picture of global climate is proving almost as difficult to produce. Scientists are confident that global warming is taking place due to emissions of 'greenhouse gases', not because they have great faith in any one of their latest models, but rather because all the different models seem to agree on this point. In the same way, predicting earthquakes and volcanic eruptions also remains a matter of estimating probabilities rather than making specific forecasts, and will almost certainly remain so. These are uncertainties that we will just have to live with.

THE EARTH'S CORE

Source of the Earth's magnetic field

The Earth has a diameter of some 12,700 kilometres, but penetrating more than about three kilometres into it is all but impossible. Geophysicists, whose speciality is the structure and composition of our own planet, cannot visit its depths to see for themselves. They have to study it by more or less indirect methods. Their chief technique is seismology, the study of earthquakes. When a major earthquake occurs, the entire planet rings like a huge bell. However, the vibrations do not travel through it evenly: their paths will be modified by regions of different density within the Earth, just as rays of light are refracted and reflected if they pass through water or a glass lens. Seismology uses the paths and speeds of the vibrations to deduce the Earth's internal structure.

By this method, Professor John Milne established in 1903 that the Earth has a thin outer 'crust' of jumbled rocks through which vibrations travel relatively slowly, and a much denser, more homogeneous inner region through which they travel much faster. (The boundary between the crust and the inner region was finally found by Andrija Mohorovicic in 1909.) However, Milne did not spot any further subdivisions within the Earth's interior. The first, and most important, of these was found by Richard Oldham in 1906. He found that slightly under half of the Earth's diameter is taken up by a 'core', whose physical properties are markedly different from the regions above.

Oldham had analysed earthquake waves in some detail. It was already known that earthquakes produce three different kinds of vibration. One kind is a surface wave that does not propagate downwards. The other two Oldham called 'first-phase' and 'second-phase' waves. He rightly suggested that the 'first-phase' waves are pressure-waves, while the 'second-phase' waves are waves of distortion of the rock through which they travel. His analyses showed that the first-phase waves travel throughout the Earth's interior as if it is homogeneous. The second-phase waves, however, do not travel through the central part of the Earth at all, or at best only very slowly. This could be explained if the Earth has a central core, with the same density as the regions above, but made of a material that is very much less rigid.

Oldham's paper was originally read at a meeting of the Geological Society of London. It cites very few references to earlier papers, but it is clear that he had discussed his findings with other members of the Society beforehand, and they had made various suggestions as to what this non-rigid core might consist of. Some had apparently even suggested that it might be made of molten iron. We now know that this is indeed the case, and we also know that electric currents flowing in it are the source of the Earth's magnetic field. In 1936 the Danish geologist Inge Lehmann largely completed our knowledge of the Earth's interior structure, by showing that right at the centre lies a small 'inner core', where the iron is compressed solid by the tremendous pressure from above.

R. D. Oldham, *Quarterly Journal, Geological Society of London*, 62, 456–475 (1906)

THE MOHO

The lower boundary of the Earth's crust

In the closing years of the nineteenth century, geologists invented instruments with which to record the vibrations produced by earthquakes, as they travelled across the Earth's surface and through its interior. The original intention was simply to study the earthquakes themselves, and in particular to pinpoint the locations of the earthquakes as they happened, anywhere in the world. However, the scientists soon found that they could use data from earthquake recordings to deduce the paths that these vibrations had traced as they spread throughout the Earth's interior, and from these they could work out the Earth's internal structure. Thus was the science of seismology born.

One of the first seismologists was the Croatian Andrija Mohorovicic, a former schoolmaster who in 1892 was appointed Head of the Meteorological Observatory at Zagreb. The Observatory's official role was in recording and forecasting the weather, but Mohorovicic had wide-ranging interests, and soon the Observatory was active in astronomy and seismology as well. He set up a network of seismic recording stations, and in 1909 they recorded data from a major earthquake centred beneath the Kupa Valley, some forty-eight kilometres south of Zagreb. The recordings clearly showed that the shock waves picked up by the more distant stations had travelled faster than those recorded nearer to the earthquake. On analysing the data closely, Mohorovicic concluded that at a level about fifty-four kilometres below the Earth's surface there is a sharply-defined boundary, with the rock below it very much denser than the rocks above. Shock waves that penetrate this boundary then travel on much faster through the denser rock below. The boundary is now known as the Mohorovicic Discontinuity, or 'Moho' for short.

Over the next few years it was found that the Moho is not just a local feature under Croatia. It is the boundary between the thin upper 'crust' of light rocks that cover the Earth's surface, and the much denser 'mantle' beneath. The existence of a light crust had been shown six years earlier by the British seismologist John Milne, but his data was not detailed enough to show how thick it was, or whether it had an abrupt lower boundary. Further studies of the Moho have shown that the crust's thickness varies. It is up to eighty kilometres thick underneath continents, but may be as little as five to eight kilometres thick under the oceans. This is no great distance, and one might surmise that it should be possible to use an offshore drilling-rig to drill right through the crust at one of its thinner points and obtain a fresh sample of rock from the mantle. However, attempts to do this have so far not succeeded. In the shallow offshore regions where drilling rigs normally operate the crust is as thick as it is under dry land. The regions where it is thinner are all under the deep oceans, where conventional drilling rigs cannot operate. The technical problems of drilling in such locations have so far proved insuperable, at least with the relatively modest budgets available to this field of research.

A. Mohorovicic, *Jahrbuch der Meteorologischen Observatorium der Zagreb*, 9, 1 (1909)

THE IONOSOPHERE

The great reflector plate in the sky

Radio is without doubt one of the greatest technical inventions of all time. From the first decade of the twentieth century, it has been used to send signals around the world, making information available to all, and enabling everyone able to speak to everyone else. However, this ability of radio to span the globe came as a surprise. Radio waves travel in straight lines, so it was originally assumed that a receiver had to be within view of the transmitter in order to receive a signal. But in 1901 the great radio pioneer Guglielmo Marconi sent a radio signal across the Atlantic Ocean, from Lizard Point in Cornwall to Signal Hill in Newfoundland, and within six years a regular radio-telegraph service was operating from Britain to the United States. Clearly radio waves could be made to go over the horizon, following the Earth's curved surface. But nobody knew how they did it.

The explanation came in 1902 from the self-taught amateur physicist Oliver Heaviside in Britain, and the electrical engineer Arthur Kennelly in America, who both independently realised that the upper reaches of the Earth's atmosphere must be acting as a reflector. In the course of their journey across the Atlantic the radio waves must be following a zigzag path, bouncing several times between the surface of the sea and the reflecting layer in the sky. This layer, originally called the 'Heaviside Layer' and now known as the 'ionosphere', reflects radio waves because the air molecules up there are electrically charged. They get their charge because they are bathed in ultraviolet light from the sun, which causes the 'photoelectric effect', as explained by Einstein in 1905. Charged materials will always reflect radio waves.

This was the theory, but how could it be proved in practice? How could the ionosphere be detected, and its height measured? The answer was found in 1924 by Edward Appleton, a physicist at King's College, London, just a few weeks after Heaviside's death. He realised that the radio waves from the BBC's transmitter in Bournemouth must reach a wireless aerial in Oxford (for instance) by two routes: a straight line, and a dogleg path bouncing off the ionosphere. When they reached his receiver, the waves that had travelled by the two routes would be 'out of step' with each other, and by measuring how far out of step they were, he could work out how far up the ionosphere must be. With the help of the BBC he did some quite simple experiments, and showed that the ionosphere must be about 96 to 112 kilometres up in the sky. Appleton was knighted in 1941 in recognition of this work.

Today we are used to having satellites to relay UHF radio signals. We may forget that for over fifty years medium and long-wave radio was used to send signals around the world relying entirely on the ionosphere to reflect them, with no need for any satellites at all.

E. V. Appleton and M. A. F. Barnett. *Nature*, 115, 333–334 (1925)

PLATE TECTONICS

Oceans and continents explained

Some discoveries are made suddenly, by one researcher, or at most a small team. Others appear more slowly. Geology in the 1960s had an abundance of facts, but it was only towards the end of the decade that they were all fitted together into a grand theory that neatly explained them all at once.

The facts are clear when one looks at a map of the world, especially if the map shows the floors of the oceans (such maps became available in the early 1960s). The following features all beg to be explained.

- The main mountain ranges of the world occur in great belts, such as that extending down the western side of North America (the Rockies) through Central America and on down the western side of South America (the Andes).
- Most of the world's volcanoes also occur in great belts. The main belt encircles the Pacific Ocean. Another runs from Indonesia north-westward to the Mediterranean. These belts are also the main earthquake zones.
- A look at the rock formations in Africa and South America shows that they were once one continent, which split apart to create the South Atlantic Ocean. North America and Europe were also once joined, as were Australia, South America, and Antarctica.

- Running all the way down the centre of the Atlantic Ocean is an undersea mountain range with a rift valley in the middle. Similar rift valleys are to be found on the floor of the Indian Ocean.

Separate explanations for all these facts were starting to appear throughout the 1960s, before they were eventually all drawn together to form the theory now called 'Plate Tectonics'. According to this theory, the Earth's surface region is divided into rigid plates, each perhaps 112 kilometres thick, which are very slowly moving. In places such as the centre of the Atlantic, two plates are moving apart, and fresh material is welling up from below to fill the gap. Elsewhere, one plate may be sliding under another (this is happening around the western edge of the Pacific Ocean); or two plates may be sliding past each other (as in California); or one plate may be ploughing into another and coming to a halt, throwing up a mountain range in the process (such as the Himalayas). Most volcanoes and earthquakes result from the friction between two plates grinding together. The power supply that drives this activity is the intense heat of the Earth's interior, which is making the surface seethe like the scum on a pan full of boiling jam, only in extreme slow motion.

The 'eureka moment' when the Plate Tectonics theory became clear was probably in 1968, when William Morgan of Princeton University published an article in the *Journal of Geophysical Research*, showing a map of the world with all the plates (he called them 'blocks') drawn on it (*see* Plate 8). It was quickly accepted by the geologists of the time, especially in Britain and Europe. The Americans and Russians came round to it more slowly, but by the mid-1970s it had become the prevailing orthodoxy.

W. J. Morgan. *Journal of Geophysical Research* 73, 1959–1982 (1968)

METHANE CLATHRATE IN THE OCEAN DEPTHS

A possible energy source for the future

Methane clathrate is extraordinary stuff. It is in essence ice, heavily impregnated with methane, better known as 'natural gas'. It is only stable at very high pressures and temperatures around freezing—it remains solid up to temperatures a few degrees above zero, but the methane fizzes out if it is not kept under high pressure. Methane is, of course, highly flammable, and so one can be faced with the strange spectacle of a block of ice burning like a fire-lighter, leaving a small pool of water behind.

Gas clathrates were first described in the early nineteenth century, but they were not found in nature until the 1960s, when oil prospectors in Outer Siberia found large deposits of methane clathrate deep in the permanently frozen soils of the tundra. Similar deposits were found a few years later in northern Alaska and Canada. The Russians tried to use the Siberian deposits as a source of natural gas, but could not overcome the technical problems. In those days ordinary natural gas was the energy source with the best future prospects, and methane clathrate was uneconomical by comparison.

Then in 1972 an American deep-sea drilling project led by Charles Hollister and John Ewing was taking echo-soundings on the Blake Ridge off the Bahamas. Deep in the mud on the seabed, there seemed to be a hard surface reflecting an echo.

When they drilled through it, large quantities of methane were given off, and they at once realised that they had found a deposit of methane clathrate. Their discovery caused little interest at the time, but during the 1980s and 1990s further exploration began to show just how widespread such deep-sea deposits are. All over the world, they have been found buried deep in the sediments at the base of the 'continental slopes', around the borders of the deep ocean basins. The methane is originally generated from the breakdown of organic matter by certain bacteria that thrive in the absence of oxygen, and in the extreme cold and high pressure of the ocean depths it slowly bonds with water to make clathrate. These deep-sea deposits may store over twice as much energy as all the other fossil fuels—coal, oil and natural gas—combined. In the closing years of the twentieth century several laboratories around the world started researching into industrially feasible ways of extracting it.

Methane is generally the cleanest of all fossil fuels, because when it is burnt it only generates carbon dioxide and water vapour. However, it is also a 'greenhouse gas', which can contribute heavily to global warming. Any attempt to extract it from methane clathrate on a commercial scale would probably result in quantities of methane being inadvertently released into the atmosphere. Some geologists suspect that at certain times during the Earth's past this may have happened naturally on a huge scale. One such event might even have caused the great 'mass extinction' of plants and animals that occurred 250 million years ago, shortly before the first dinosaurs appeared.

C. D. Hollister et al, *Initial Reports of the Deep Sea Drilling Projects*, XI, U. S. Government Printing Office (1972).

BLACK SMOKERS

Super-hot geysers on the ocean floor harbouring exotic life forms

In some ways we know less about the surface of the Earth than we do about the surfaces of the Moon or Mars. Four-fifths of it is covered with water, most of it to a depth of more than three kilometres, and what lies underneath remains almost completely hidden from view. In the last three decades of the twentieth century research on the ocean floors intensified, much of it in order to test the Plate Tectonic theory of geology. This research was made possible by the use of deep-sea submarines, most notably the *Alvin*, owned by the Woods Hole Oceanographic Institute in Massachusetts, USA. To the general public, *Alvin* and its chief pilot Robert Ballard are famous for exploring shipwrecks such as the *Titanic* and the *Bismarck*. However, to scientists Ballard's greatest achievement was probably his part in the discovery of the 'black smokers', perhaps the strangest ecosystems to be found anywhere on Earth.

In 1976 Ballard and his submarine joined a team exploring an area of the Pacific Ocean off the coast of Mexico, some eighty kilometres south of the tip of the Baja California peninsula. The main feature of the ocean floor in this region is a ridge which, according to Plate Tectonic theory, marks the border between two of the great rigid 'plates' that make up the Earth's surface. These two plates are slowly moving away from each other, so fresh rock, possibly in the form of molten lava, is expected to be constantly welling up to fill the opening gap between them. Over a period of three years, the team explored the geology of the ridge, and in the process found several geysers or 'hydrothermal vents' spewing water black with suspended mineral particles, mostly sulphides. In some of these, temperatures of over 300 °C were recorded. The immense pressure found at such depths prevented the water from boiling even at these temperatures, so it rose from the seabed like dense plumes of smoke. The sulphide particles precipitated out of the water jets, forming cones and even chimney-like structures many metres high.

These black smokers have since been studied as intensively as is feasible, given their inaccessibility. Extraordinarily, they actually harbour life. Within the hot-water plumes are found microscopic organisms related to bacteria, but sufficiently different that a new name has been coined for them—the Archaea. In the water immediately around the vents live shrimps and other animals, all of them unique to this kind of habitat. The Archaea themselves have generated great interest. They seem in many ways to be the most primitive of all living things, having diverged from bacteria billions of years ago. This has led to renewed speculation among biologists about how life began on Earth. Previously it had been assumed that the first living things inhabited shallow water, perhaps along sea shores, where they were warmed by sunlight. The discovery of the Archaea suggested that this may not have been the case after all—rather, life may have originally have been warmed by heat from geysers.

F. N. Spiess et al, *Science*, 207, 1421–1433 (1980)

THE OZONE HOLE

Gas from aerosol cans destroys our shield from ultraviolet radiation

The ozone layer was discovered in the early twentieth century. Some nineteen kilometres above us, ultraviolet light from the sun causes oxygen in the upper atmosphere to turn into ozone. This gas absorbs the ultraviolet light, preventing much of it from reaching ground level. This is just as well, because ultraviolet is hazardous: it causes skin cancer in humans, and stunts the growth of plants, among other undesirable effects. However, ozone is an unstable compound, and it is constantly breaking down into oxygen as fast as it is being generated. This means that the ozone layer exists permanently in a fragile equilibrium, building up to a concentration where it breaks down at precisely the same rate as it is being generated.

In the 1970s, atmospheric chemists started to worry about the effects of air pollution on the ozone layer. One of the pollutants that particularly concerned them was chlorine. Chlorine levels in the upper atmosphere were predicted to rise, with the increasing use of chlorofluorocarbons (CFCs) in industry. These gases had been used in refrigerators ever since the 1950s, and were also becoming widely used as propellants in aerosol cans and fire extinguishers, and in the manufacture of plastic foams. They were popular with industry because they were highly stable and inert, and therefore apparently safe. However, they do break down slowly, and as they do so, they release chlorine gas. Chlorine, being light, tends to drift upwards and accumulate in the upper atmosphere. There, it triggers the chemical breakdown of ozone to oxygen.

In 1985, a team of British scientists working at a research station in Antarctica first reported that for much of the year, the ozone overhead largely disappeared. It was soon found that the loss of ozone was most marked in regions where chlorine was most abundant. The prevailing winds in the upper atmosphere were blowing the chlorine towards the Arctic and Antarctic, where it was accumulating and destroying the ozone. The Antarctic was most severely affected, but a marked loss of ozone was also found to be taking place over and around the Arctic. This area of ozone depletion was spreading year by year, and was soon encroaching on densely populated parts of Europe and North America.

Generally it has proved very difficult for all nations to agree on any common policy on environmental matters. But when faced with the threat of dangerous ultraviolet radiation penetrating the atmosphere to ground level, they acted fast. In 1987 most of the industrialised nations signed the Montreal Protocol, which committed them to an immediate freeze and subsequent 50 percent reduction in CFC output. Later amendments bound them to even greater reductions.

Critics sometimes ask why governments spend money on apparently useless academic research, such as studies of the atmosphere over Antarctica. On this occasion, however, the research justified itself. The ozone hole was discovered by chance, in the course of a long-term data-gathering project originally started in the 1950s. It may have averted a major environmental disaster.

J. C. Farman et al, *Nature*, 315, 207–210 (1985)

CHEMISTRY

One sometimes hears people say that the nineteenth century was the really exciting age of chemistry. It was indeed a period of great advances, with two in particular standing out from the rest: the atomic theory, and the 'Periodic Table'. The atomic theory, first advanced by John Dalton around 1803, states that all matter is made of atoms, with each chemical element having atoms of a specific weight and 'valency'. (An atom's valency is, roughly speaking, the number of other atoms it can bond with when forming a molecule.) The 'Periodic Table' of the chemical elements, first published by Dmitry Mendeleev in 1869, groups the different elements together in families, according to their valency and general chemical properties. However, Mendeleev had no idea as to why the elements could be grouped in this way. These two scientific milestones laid the foundations of chemistry as it has been practised ever since. By the middle of the century chemists knew enough about the properties of different elements and compounds that they could start synthesising new substances with particular required properties. Industrial chemistry as we know it today, in which commercial companies maintain laboratories where new compounds are devised to order, originated in the 1850s with the invention of the first synthetic dyes and plastics.

However, the chemistry of the nineteenth century was really as much a craft as a science. Beyond Dalton's atoms, it had very little by way of underlying theory. Chemists simply learnt from experience what worked and what did not, without having very much idea why. The big achievement of the chemists of the twentieth century was to give their subject a true theoretical basis. This only became possible as a result of the advances made in physics from the 1890s onwards. In particular, physicists finally proved to everyone's satisfaction that atoms really exist; showed that each chemical element has atoms with a specific number of protons, neutrons and electrons; and showed that the electrons occupy discrete 'orbitals' around an atom's nucleus. On the basis of this knowledge, chemists were at last able to produce a theoretical explanation for everything that they had discovered to date. According to this theory, the electrons' orbitals form a series of concentric shells, and an element's chemical properties are largely determined by two factors: the weight of its atoms, and the number of electrons each atom has in its outermost electron shell. Chemistry in the twentieth century thus became a true science, capable of making predictions based on theory, in a way that nineteenth-century chemistry was not.

However, this does not mean that modern chemists no longer resort to trial and error. The hunt for 'high-temperature' superconductors, to take an obvious example, is still proceeding by little more than informed guesswork—nobody yet knows how they really work, so nobody can predict at all accurately what makes a compound able to superconduct at a given temperature. Two of the other discoveries described in this section, Ziegler-Natta catalysts and buckminsterfullerene, were also made quite by accident, to the considerable surprise of the scientists involved, who were working on other questions at the time. Chemical theory may no longer be a closed book, but it is far from being an open-and-shut case.

This section is a short one, but not because little has happened in chemistry in the last century—far from it. There are two main reasons why the section is not longer. The first is that many of the really notable advances in chemistry, from the viewpoint of non-chemists at least, have been inventions rather than discoveries. Chief among these have been the plastics that are now so familiar, only a handful of which were in existence in 1900 (indeed, most of them have only been manufactured in large quantities since the 1950s). In addition there are modern glasses and ceramics such as those used in heat-proof kitchenware, the whole technology for building miniature electronic circuits on silicon chips, and many other innovations that have transformed our world. All these fall outside the scope of a book about discoveries. Some of the advances that have been included—the Haber Process, high-temperature superconductors and Ziegler-Natta catalysts—could also be described equally as discoveries or as inventions. Chemistry has always been an intensely practical science.

The other reason why this section is so short is that many of the really exciting chemical discoveries have been in the fields of biochemistry, molecular biology, and pharmaceutics. The Medicine, Biology and Genetics sections of this book include accounts of such discoveries as sulphonamide drugs; the isolation of urease; the Krebs, Calvin and ATP cycles; the composition of insulin; and the structures of the DNA and Vitamin B12 molecules. The scientists involved may have styled themselves biologists, but what they were really doing was chemistry under a different name.

Inevitably with its emphasis on practical and industrial applications, chemistry attracts controversy. Nowhere is this more clearly illustrated than in the career of Fritz Haber, the scientist whose key discovery, the Haber Process, simultaneously fended off the prospect of global famine for at least a generation and made possible the mass carnage of the two World Wars. He also became notorious for his work in the development of chemical weapons. When most people think of the chemical industry as a whole, they probably think at least as much about the pollution it causes as the commercial products that have so transformed our lives. Chemistry has penetrated the whole of the manufacturing, farming and healthcare industries. It deserves full credit for their positive achievements, but in many people's eyes its successes will always be tainted by the ill-effects that go with them.

THE HABER PROCESS

Cheap ammonia changes the course of history

Nitrogen is one of the most plentiful of all chemical elements, making up about four-fifths of the air we breathe. So it may seem strange that until 1909 it was actually in short supply for industrial use. The demand for it is enormous, especially for the manufacture of nitrate fertilisers. Traditionally, the main fertilisers used throughout the world have always been animal and human dung, which are rich sources of nitrogen compounds. However, in the nineteenth century the industrialised nations found these quite insufficient, and other sources of nitrates were needed. The only natural sources were 'chile saltpetre' (the mineral potassium nitrate), and 'guano'—the accumulated droppings of the millions of sea-birds that breed on islands off the coast of Chile. These were not nearly enough.

A method was needed to extract nitrogen from the air and turn it into nitrates. However, pure nitrogen is very unreactive (which is why there is so much of it). This was the problem solved by the German chemist Fritz Haber, who found he could make ammonia from nitrogen and hydrogen, by mixing them at a pressure of 200 atmospheres (20 megapascals) and a temperature of 600 °C, and using osmium as a catalyst. Ammonia is a very reactive gas, and it is easy to make nitrate compounds from it. Haber's original method only produced ammonia in small quantities, but he further developed it into the Haber-

Bosch process, which uses iron filings as a catalyst instead of the original osmium. It remains the main industrial method to this day.

Cheap nitrate fertilisers have been the basis of the agricultural revolution, which coincided in the twentieth century with a massive increase in the human population. In 1900, there were some two billion people on earth. A hundred years later, the world population was six billion and still rising fast, but global starvation had still not occurred. However, intensive farming methods were causing severe environmental degradation in many regions. Meanwhile, nitrogen compounds were also being used to make explosives, and it may fairly be said that the Haber process was not only at the root of twentieth-century agriculture, but also of twentieth-century warfare. No other single chemical reaction can have changed the world so much, for better and for worse.

Fritz Haber was a very public-spirited man, who developed his chemical process with the explicit aim of benefiting humankind. However, he was also intensely patriotic. During the First World War he willingly helped in the development of chemical weapons, despite his revulsion for the suffering they caused, because he hoped they would deliver a quick victory to the Germans. His wife was so upset that she committed suicide. When Germany lost the war and had to pay huge reparations, Haber tried unsuccessfully to find an economic way to extract gold from sea-water, to help pay off the debt. But when Hitler came to power, his patriotism was of no account, because he was Jewish. He left Germany in 1933 with the intention of settling in Britain, but died the following year in Switzerland.

F. Haber, *Chemiker Zeitung,* 34, 245–246; *Journal für Gasbeleuchtung,* 53, 367–368 (1909)

CHEMICAL BONDS

How molecules are held together

The science of chemistry is the study of how atoms of the different chemical elements bind to each other to form all the different compounds we know. So it may seem surprising that the most exciting advances in chemistry were made in the nineteenth century, when no-one had the faintest idea of how atoms bind to each other, and indeed many scientists doubted that atoms existed at all. However, in the early years of the twentieth century the existence of atoms was finally proved, and it was shown that an atom consists of a nucleus surrounded by electrons. Immediately chemists realised that the electrons must hold the key to the problem of how elements bond with each other to form compounds. The classic paper that first set out our present ideas about chemical bonds was published in 1916 by Gilbert Lewis, professor of chemistry at Berkeley, California.

Lewis' big idea was that an atom's electrons are arranged in concentric shells, and all of chemistry follows from the need to fill an atom's outermost shell with its maximum quota of electrons. The elements hydrogen and helium have just one shell, which can contain a maximum of two electrons. All the other elements have more shells, commonly containing up to eight electrons each. The 'inert gases'—helium, neon, argon, krypton and xenon—all contain a complete complement of electrons in the outermost shell, which is why they generally do not react to form compounds with other elements. All other elements have gaps in the outermost shell. Some non-metallic elements, especially carbon, will fill these gaps by sharing electrons with other atoms (this is now called 'covalent bonding'). Typical covalent compounds are hydrocarbons like butane gas. Metallic elements, meanwhile, will simply give away their outermost electrons to non-metals; for instance, sodium gives an electron to chlorine to make common salt, sodium chloride. Such compounds are called 'ionic' (Lewis called them 'polar', but this word has another meaning in modern chemistry). In a particular stroke of genius, Lewis saw that most of chemistry could be explained if there is a continuum of compounds, with ionic ones like salt at one end, covalent ones like butane at the other end, and most other compounds somewhere in between.

It is hardly an exaggeration to say that all of today's basic chemistry, as one might learn at pre-university level, follows from Lewis' insight. However, Lewis was completely wrong about one thing. He believed that atoms were cube-shaped, and the electrons sat stationary at the corners of the cube. He had heard of Rutherford's theory that electrons orbit round the nucleus, and even of Bohr's quantum version, which puts the electrons into discrete 'orbitals', but he dismissed these out of hand, saying they were 'logically objectionable'. It was not until the early 1930s that another great chemist, Linus Pauling, showed that atoms behave as Lewis described precisely because Bohr's model is right. Pauling got a Nobel Prize, but he always gave Lewis full credit for getting the basic idea first.

G. N. Lewis, *Journal of the American Chemical Society,* 38, 762–785 (1916)

POLYMERS

What are plastics? The biggest molecules ever

Until the late nineteenth century, the basic materials for all building and manufacture had remained the same throughout history: stone, ceramics and glass, natural fibres for textiles, metals, wood and leather. The nearest thing to plastics as we know them were amber, lacquer, and india-rubber. These were used only in very small quantities. Then in 1862 celluloid was invented, and within a few decades was being used to make photographic film and cheap stiff collars. By the time of the First World War it had been joined by rayon, vulcanite (used for telephones, electric plugs and fountain-pens), and bakelite (used for the knobs on cars' dashboards and picnic crockery). Several other plastics, including pvc (polyvinyl chloride) were also invented in the nineteenth century, but nobody could devise a means to manufacture them in bulk.

All these early plastics had been invented largely by trial and error. The underlying chemistry was not understood at all. Chemists realised that the individual molecules of these plastics must be larger than those of the raw materials from which they were made, but they had no idea how much larger. They imagined that each molecule of celluloid, for instance, must consist of just a few molecules of cellulose bundled loosely together. However, a few chemists were not convinced by this model. Among them was Hermann Staudinger at the Zurich Polytechnic, who is now acknowledged as the founding father of the plastics revolution. In a series of experiments on india-rubber, he was driven to the conclusion that it consists of enormous chain-like molecules, far bigger than anyone had ever envisaged before. He called these molecules 'polymers'. Each link in the chain was a molecule of isoprene, which consists of just five carbon atoms and eight hydrogen atoms. Staudinger devised a means of measuring the size of these huge molecules, by measuring their viscosity when they were molten. He found that the size of a polymer molecule could vary enormously, but it could typically be made of thousands or even tens of thousands of small molecules linked together.

Staudinger's fellow academics were quite dismissive of his ideas to start with. However, within the growing plastics industry they generated interest, which rapidly increased when he started inventing new plastics, including polystyrene. No longer did their chemists have to rely on guesswork, there was now a solid theory to guide their research. In America especially the use of plastics started to spread. In 1928 the Du Pont corporation recruited Wallace Carothers from Harvard University, who developed Staudinger's theory and invented nylon and several other plastics before his early death nine years later. Carothers realised that polymer molecules did not have to be straight chains: some could be huge tangles of side-branches.

Finally after the Second World War, Professors Karl Ziegler and Giulio Natta pioneered the use of catalysts to make organic molecules form polymers to order. Using this means, the number of different plastics available suddenly increased enormously, and the modern plastics industry was born.

H. Staudinger and J. Fritschi, *Helvetica Chimica Acta*, 5, 785–806 (1922)

PLUTONIUM

The power behind the Bomb

During the twentieth century, several new chemical elements were described. Most of these were 'trans-uranium' elements, highly radioactive and unstable, which were made in minute quantities in nuclear reactions. Their discovery caused some interest among atomic scientists, but no great stir. One of them, however, had some of the most far-reaching repercussions of any scientific discovery. This element was plutonium.

By 1940, teams of scientists in both Britain and America (including many expatriate Europeans) were working on nuclear energy. Similar work was also being done in Nazi Germany, but the researchers there were not as highly organised, and they made very little progress. Two possibilities were being considered. One was the 'nuclear pile', in which a quantity of uranium was so confined that a controlled chain reaction took place: as its atoms underwent radioactive decay and disintegrated, some of the fragments would hit other atoms and cause them to disintegrate in turn. The other possibility was the atomic bomb, in which conditions are created for a runaway chain reaction. The problem with building a bomb is that normally, uranium is not radioactive enough for a runaway reaction to occur. Uranium occurs as a mixture of two 'isotopes', differing in the number of neutrons in each atom. The more abundant isotope, uranium-238, is only mildly radioactive. The much less abundant uranium-235 is radioactive enough, but separating it out from uranium-238 is very difficult and costly.

However, in the spring of 1940, a physicist at Princeton University called Louis Turner suggested that if a nuclear pile were successfully built, one of the products of the chain reaction inside it would be a new stable element, heavier than uranium, and radioactive. A few weeks later, this element was observed by Edwin McMillan and Philip Abelson, two researchers at Berkeley University in California. They bombarded a small sample of uranium with neutrons, not in a nuclear pile, but using a 'cyclotron' particle accelerator. They found that it was converted into an unstable element (now called neptunium), which rapidly decayed to form the predicted new stable element, plutonium.

A few facts about plutonium rapidly became apparent. Firstly, it was radioactive enough for a runaway reaction to be possible. Secondly, any controlled reaction in a nuclear pile would inevitably generate plutonium as a by-product. Indeed, it was much easier to make plutonium from uranium than it was to separate uranium-235 from uranium-238. And lastly, the 'critical mass' of plutonium needed for a runaway reaction was much less than that of uranium-235. A plutonium bomb would thus be smaller and cheaper than a uranium bomb. It followed that any nuclear power programme would inevitably produce the materials for building bombs, and the peaceful and military uses of nuclear energy could not be kept separate.

In August 1945, a uranium bomb was exploded over Hiroshima, and a plutonium bomb was dropped on Nagasaki. The two bombs proved to be equally destructive, and so the cheaper plutonium was to become the main raw material of the world's nuclear arsenals.

E. M. McMillan and P. H. Abelson, *Physical Review*, 57, 1185–1186 (1940)

ZIEGLER-NATTA CATALYSTS

Chance discovery leads to cheaper and stronger plastics

Devotees of old movies will know *The Man in the White Suit*, a much-loved comedy in which a chemist (played by Alec Guinness) invents a new synthetic fibre entirely by accident. Far from being complete whimsy, such chance breakthroughs do indeed happen. One of them occurred in 1952 in the laboratory of Karl Ziegler at the Max Planck Institute for Coal Research at Mulheim, Germany, and led to the sudden arrival of polythene as the ubiquitous lightweight plastic.

Polythene was not entirely new. It had been invented, again completely by accident, by chemists at ICI in 1933. However, their early version was not as strong as the polythene we know today, and moreover its manufacture required immensely high pressures, which made it expensive. But then in late 1952, one E. Holzkamp, a graduate student in Ziegler's laboratory, was trying to synthesise hydrocarbons with an aluminium atom attached. He started with aluminium triethyl (a molecule consisting of six carbon atoms, fifteen hydrogen atoms and an aluminium atom), and was trying to add more carbon and hydrogen to it by reacting it with the gas ethylene (two carbons and four hydrogens). The experiment did not work. The ethylene molecules, instead of linking up with the aluminium triethyl, linked up with each other in pairs. Painstaking detective work showed that this effect was somehow being caused by minute amounts of nickel left in the reactor chamber from a previous experiment.

Ziegler guessed at once that his students could be onto something big. If the ethylene molecules would form pairs in this way, might they be persuaded to form long chains, making polythene? Over the next two years, the team found that indeed they could. It was best done by using a mixture of aluminium triethyl and titanium chloride (this worked better than nickel), at temperatures of 50–100 °C and—to everyone's immense surprise—ordinary atmospheric pressure. Moreover, the polythene they made this way was much stronger than that produced by ICI's high-pressure method. We now know that this is because it is 'stereospecific'—the molecules are straight chains of carbon atoms, rather than the randomly-branched chains made by the ICI method.

Over the next few years, it was found that mixtures of organic aluminium compounds and titanium or similar metals could be used as catalysts to produce many different kinds of plastic, always with stereospecific molecules. These mixtures are called 'Ziegler-Natta catalysts', after Ziegler and Giulio Natta, an Italian chemist who further developed their use, and invented polypropylene, the plastic commonly used for making lightweight rope.

Ironically, it later appeared that Ziegler's team were not the first chemists to make polythene by this means. In 1943, Max Fischer at the BASF laboratories was using a very similar reaction to make synthetic lubricating oils from ethylene. He noticed that in addition to oil, the reaction always produced a large amount of white powder. Cheerfully unaware of what it was, he just threw it away as an unwanted contaminant, thus missing a Nobel Prize and a place in the history of chemistry.

K. Ziegler et al, *Angewandte Chemie*, 67, 541–547 (1955)

BUCKMINSTERFULLERENE

The soccer-ball molecule

Of all the elements, carbon has the most diverse chemistry. This stems from the unique ability of carbon atoms to link up in chains and lattices of any degree of size and complexity. Pure carbon has long been known to occur in two forms: diamond, where the carbon atoms form a lattice with each atom linked to four others; and graphite, where each atom links to three others, in a flat lattice of hexagons like chicken-wire. Then in the late 1980s, a third configuration was found: 'buckminsterfullerene', the molecule that looks like a football.

The new molecule was discovered quite by accident. Harry Kroto was (and still is) a chemist at Sussex University. In 1985 he was working at Rice University at Houston, Texas, with some chemists there who were interested in the large carbon molecules that astronomers were finding drifting in outer space. They surmised that these originated in old stars, where carbon is formed by nuclear fusion, and were spread through outer space when these stars exploded as 'supernovae'. They devised a bench-top experiment to simulate the effects of a supernova, by holding a lump of graphite under a flow of helium gas, and blasting it with a powerful laser. As expected, the laser vaporised carbon molecules of various sizes off the graphite. However, there was a surprise: instead of getting the broad range of molecules they expected, they found that molecules with just 60 atoms were much more abundant than the rest. Clearly a 60-atom molecule must be more stable than anything larger or smaller.

Harry Kroto has always had a keen interest in design and architecture. There came into his mind the image of a soccer ball—a polyhedron with 20 hexagonal sides (traditionally coloured white), 12 pentagonal sides (coloured black), and just 60 corners where the seams meet. Maybe the carbon atoms were forming a polyhedron like a tiny soccer ball with an atom on each corner. Kroto and his colleagues named the molecule buckminsterfullerene (or 'buckyball' for short), in memory of the architect Buckminster Fuller, who designed light-weight domes supported by hexagonal lattice frameworks. It was only later that they learnt that two Japanese chemists, Z. Yoshida and E. Osawa, had predicted this molecule's existence some fourteen years earlier. Unfortunately, they had published their paper in Japanese, which neither Kroto nor his colleagues could read.

It was five years before methods were devised to synthesise buckminsterfullerene in large enough quantities for its chemistry to be studied, and Kroto's hypothesis to be proved right. As yet it is too early to say whether it will have any applications, but it appears to have interesting properties as a semiconductor (especially if a metal atom is caged inside it), and it may also have medical uses. Its discovery has also opened up an exciting new field of carbon chemistry. A whole range of spherical molecules ('fullerenes') and tubular molecules ('nanotubes') can now be made. Some of these may be used in future to make new extremely strong lightweight materials.

H. W. Kroto et al, *Nature*, 318, 162–163 (1985)

HIGH-TEMPERATURE SUPERCONDUCTORS

Promise of a revolution in electrical engineering

Superconductivity—the ability of certain substances to lose all resistance to electric current when cooled to sufficiently low temperatures—sounds like an engineer's dream come true. However, in the ninety-odd years since its discovery in 1911 it has only found a very few specialist applications, mostly in laboratory apparatus. Practical difficulties have prevented it from being used widely. The temperatures required are only a few degrees above the absolute zero of 273 °C, and the only practical way to keep electric circuits at such low temperatures is to immerse them in liquid helium, which is very expensive and difficult to handle. So for three-quarters of a century after its discovery, superconductivity seemed to be little more than a curiosity, of interest to physicists who wanted to know how it worked, but seldom to anyone else.

Then in 1986 came the breakthrough. Alex Muller and Georg Bednorz, working at the IBM research laboratory at Ruschlikon in Switzerland, had become interested in a class of ceramics called perovskites. They knew that an oxide of lithium and titanium had been shown to become a superconductor at 260 °C, which was then considered almost hot by superconductor standards. They suspected that perovskites, which are complex oxides of several metallic and non-metallic elements, might possibly become superconductors at even higher temperatures. So they synthesised some hundreds of different compounds, and tested each in turn. Finally they announced that an oxide of copper, lanthanum and barium became a superconductor at the unheard-of temperature of 243 °C, a full twelve degrees higher than the previous record. Soon afterwards they found that if the compound contained a little lead as well, it would superconduct at 215 °C.

To most people, these might still seem like extremely low temperatures. But in fact, they are not far below the point at which superconductors could be used in industry. Any substance that superconducts at 196 °C could have commercial potential, because that is the temperature of liquid nitrogen, which is produced in bulk and is extremely cheap. So Muller's and Bednorz's discovery created immense excitement, and for several months hardly a week went by without some laboratory claiming (often very prematurely) to have broken previous temperature records for superconductors. At the time of writing, the record stands at 135 °C, and cables made of superconducting tape wound round a tube filled with liquid nitrogen have been installed in the power grids of Copenhagen and Detroit. A major problem lies in the extremely brittle nature of most of these synthetic compounds, which makes it very difficult to fashion wires or electrical components from them.

Nobody yet knows quite how these 'high-temperature superconductors' work. In conventional low-temperature superconductors, the electrons pair up to form 'Cooper pairs', which lack the 'spin' that all single electrons have, and can bounce past the atoms of a substance without losing any of their energy on the way. However, Cooper pairs can only form at extremely low temperatures, so in high-temperature superconductors some other mechanism must be involved.

A. Muller and G. Bednorz, *Zeitschrift fur Physik B: Condensed Matter*, 64, 189–193 (1986)

PHYSICS

For physicists, the twentieth century, and especially its first half, was the heroic age. In 1900, they generally believed that they knew almost everything, and the end of their subject was in sight. Within five years, they realised that they knew practically nothing, and that their science was only just beginning. By the century's end, the cycle was, just possibly, about to repeat itself. Physicists today are largely confident that their model of how the universe works is the right one as far as it goes, but there is also an uneasy realisation that when they start to fill in the remaining gaps they may be in for some very big surprises.

Modern physics may conveniently be said to have begun in 1900, with Max Planck's quantum theory. Before Planck there was the 'classical physics' of Galileo and Newton, which described the workings of objects no smaller than an atom, and no larger than, say, the Solar System. In the nineteenth century classical physics expanded to include electricity within its realm of understanding, culminating in the discovery of the electron, electricity's basic unit, in the 1890s. This might have alerted physicists to the depth of their ignorance. At a time when they were still arguing over whether atoms really existed or not, here was a particle which was apparently one of the atom's constituent parts. Meanwhile, the discovery of radioactivity, also in the 1890s, was a clear sign that physics still had a very long way to go. But it was only after Planck that physicists fully acknowledged the fact.

The chief characteristic of modern physics is that it is counter-intuitive. One can visualise the old physical laws of Newton or Faraday by thinking of everyday objects; but the new physics of Planck, Einstein, Bohr and their colleagues can only be described by mathematical equations written down on paper. We cannot easily imagine how something can be simultaneously both a wave and a particle, or how space and time are distorted by gravity, or how a particle has neither a fixed position in space nor a set velocity until one or the other (never both) is measured. Our minds can cope with the behaviour of objects large enough to be seen with the naked eye, but no larger than a large planet. Modern physics tells us that on scales larger or smaller than this, our intuition simply gives the wrong answers.

Physics is, on the face of it, the purest and most abstruse of sciences. And yet the achievements of its theorists in the first half of the twentieth century had further-reaching consequences for society at large than any other branch of science. They made possible two of the century's defining technical achievements: the hydrogen bomb and the silicon chip. It has often been remarked that the most destructive weapons ever made owe their existence originally to a small number of pacifist intellectuals. (Einstein is often mentioned in this context, although in fact his contribution was marginal.) The paradox largely disappears when one realises that many of the theoretical physicists were Jewish. They knew that Nazi Germany had a nuclear bomb project, and they reasoned that if Hitler got the bomb ahead of anyone else, nothing could stop him from ruling the world. It was only after the War was over that they learnt that the German nuclear project had got nowhere, but by then it was too late to do anything about it. Nuclear weapons could not be dis-invented. So it becomes less strange to see how some of the strongest advocates of nuclear disarmament have been scientists who had helped in making nuclear weapons possible in the first place.

Happily twentieth-century physics' other major product, the transistor and the integrated circuit, is more benign. Like so many technical advances, it has tended to benefit the already-rich and powerful more than anyone else, but its chief effect has been to enable people to store and spread information in far greater quantities than ever before. If Alexander Solzhenytsin were right when he said that 'the salvation of mankind lies only in making everything the concern of all', then information must be the world's most vital commodity after drinking-water, and any device that helps in handling it should surely be welcomed by everyone. In addition, practically every mechanical device one can think of now contains a chip somewhere, to improve its performance.

The bomb and the chip must help to explain why the governments of the Western world are willing to give such lavish funding to physics research today. As this book is being written, CERN, the Centre Européenne de la Recherche Nucléaire, near Geneva, Switzerland, is replacing the world's most powerful particle accelerator with one even more powerful still, at an estimated cost of some £1.6 billion, paid for by the governments of several European countries, including Britain. At first glance, one cannot imagine that the knowledge it will lead to could have any practical applications, but then people thought the same of Rutherford's experiments in the early twentieth century, and were proved wrong. More immediately, CERN produces physicists, many of whom may then go on to apply their training and expertise to more immediately practical purposes. As an obvious example, it was a CERN physicist, Tim Berners-Lee, who originally devised the World Wide Web. If Europe were to scrap the laboratory, many of the professors and students who currently work there might emigrate to the USA. There is a clear rivalry between the two sides of the North

Atlantic to have the biggest and best-equipped physics laboratories, and so (hopefully) attract the ablest researchers. Doubtless the governments involved are also watching closely to see if any of the latest advances might have military applications.

However, one does not have to think of the applications of modern physics in order to be impressed by it. The humorist Max Beerbohm once wrote that he did not really understand Einstein's General Relativity theory, 'but I do know a good thing when I see one'. The many people who have read, but probably not fully understood, Stephen Hawking's books may well agree. Physics asks the most fundamental questions about how the universe works, and the stranger the answers it finds, the more awe-inspiring they can be.

THE QUANTUM THEORY

Quiet beginnings of a revolution

In 1900 most physicists believed that their job was nearly finished. The main principles were known, they thought, and all that needed to be done was to tie up a few loose ends. They could not have been more wrong. Physics was only just about to begin.

The man whose work triggered the revolution was himself no revolutionary. Max Planck had even said that he had no wish to make new discoveries. In 1900, at the age of 42, he had set himself the task of studying how objects glow when heated. (In their usual confusing way, physicists call this 'black-body radiation'.) We all know that when a steel poker, say, is heated to a certain level, it starts to glow dark red. If it is heated further it glows orange, then yellow, and finally white. What is happening is that as the temperature is raised, higher and higher frequencies of light are being added to the glow, starting in the infra-red, moving through the red end of the spectrum and up through orange, yellow, green, blue, violet, and on into the ultra-violet. Planck's aim was to derive a mathematical formula that would connect the range of frequencies in the glow with the temperature needed to produce it.

It took him just eight weeks of extremely hard work to come up with a solution, but it required two huge mental adjustments. First, he had to assume that matter was made of atoms. In 1900 there were still scientists who did not believe this, and until then Planck had been one of them. Then came the big step. An atom cannot have any heat you want: as you heat it up, the amount of energy it possesses goes up in steps. Just as a piano cannot play a sliding scale like a trombone, so an atom can only have a range of discrete levels of energy. In order to shed energy and cool down, an atom has to 'go down the scale' step by step, giving off light at a specific frequency for each step. These tiny steps appear in Planck's formula in the form of a constant quantity, always known as Planck's Constant, and written as the letter *h*. It seems to be one of the few fundamental constants on which the entire workings of the universe rest.

All the discoveries made in the twentieth century about matter and energy rest on Planck's Quantum Theory—the knowledge that energy comes in discrete packets, always known as 'quanta'. However, Planck himself played little further part in the ensuing revolution of knowledge. It was others, starting with Einstein, who produced equations linking energy, matter, time and space, and so showed that if energy came in quanta, then everything else must do so too. Wherever one looks in the physicists' equations, Planck's Constant appears again and again.

Meanwhile, ordinary people have started using the phrase 'quantum leap', to mean a very large change in something, quite failing to realise that to scientists it means the opposite—the smallest possible change in anything.

M. Planck, *Verhandlungen der Deutschen Physikalischen Gesellschaft*, 2, 237 (1900)

RADIOACTIVE DECAY

Transmutation of elements

In the closing years of the nineteenth century, a young farmer's son from New Zealand, who was to become one of the greatest experimenters of all time, arrived in Cambridge. His name was Ernest Rutherford. In a very few years he became a Fellow of the Royal Society, and went to McGill University in Canada to become Professor of Physics there. This remote backwoods college, as it was then, just happened to have one of the best-equipped physics laboratories in the world. (By today's standards, of course, it was amazingly primitive. *See* Plate 4.) There, he set to work studying the newly-discovered phenomenon of radioactivity.

In the 1890s scientists had discovered that some elements, such as uranium, thorium, polonium and radium, emitted invisible rays that could darken photographic plates exposed to them, just as light does. The only sources of heat and light then known were chemical reactions, so it was at first assumed that these elements must be undergoing some kind of spontaneous chemical process. However, by 1900 it had already been shown that pure uranium produced radiation. Rutherford studied the radioactive ore pitchblende (which contains several radioactive elements, including uranium and radium), and pronounced that the radiation coming from it could not possibly come from chemical reactions. There was too much energy being given off, while the composition of the pitchblende was left unchanged, as far as anyone could detect. Rutherford was beginning to think that the radiation must be coming from inside the atoms themselves. This was heresy. Most scientists believed in the existence of atoms, but assumed that these were the ultimate, indivisible building-blocks of matter. The idea of disintegrating atoms was thought to be a contradiction in terms.

In 1901 Rutherford was joined at McGill by Frederick Soddy, a young chemist recently graduated from Oxford, who was initially one of his fiercest critics. However, they got to work together, using the mildly radioactive element thorium as their material. Rutherford showed that the thorium was slowly converting itself into other elements: first it turned into an unidentified substance he called 'thorium-X', which then converted itself over the space of a few days into a gas (now called thoron), that itself turned into helium over a period of a few minutes. The thoron caused any substance it touched briefly to become radioactive, too. Soddy, meanwhile, did the chemical analyses to show that these substances were each separate elements that were being generated out of the thorium. As the results of their experiments became ever more clear, Soddy was forced to concede that Rutherford was right. 'This is transmutation!' he exclaimed at one point. Nobody had thought this was possible since the days of the alchemists in the sixteenth century. But beyond doubt, they were seeing atoms of one element spontaneously disintegrating to make atoms of other elements. Radioactivity was not a by-product of this process: it was the process. Rutherford and Soddy had provided the best evidence yet that the supposedly indivisible atoms exist—by observing them falling apart.

E. Rutherford and F. Soddy, *Philosophical Magazine*, 5, 455–457(1903)

PROOF THAT ATOMS EXIST

Einstein's debut

In 1905, there suddenly appeared the man whose name has become synonymous with scientific genius: Albert Einstein. Aged twenty-six, and working in a humble job in the Zurich Patent Office, he produced three of the most breathtaking papers that physicists had ever seen. First he showed that Brownian motion was proof of the existence of atoms. Then he showed that light was composed of particles. Finally he produced the Special Theory of Relativity. Each time, he produced a dazzlingly clever solution to a well-known problem, simply by arguing from first principles. Each paper covers just one side of a page, and cites no earlier references. They are among the most extraordinary intellectual achievements of all time.

The first of these three papers tackled an old mystery. It is possible to look at smoke, for example, through a microscope, and see that it is made of tiny particles. When one sees them, the first thing one notices is that they are in constant motion. They are forever jiggling at high speed, but staying in much the same spot from one second to the next. This restless movement is called Brownian motion, after its discoverer, Richard Brown in the 1820s. What causes it? A few people had suggested that the smoke particles were being jostled by molecules of air, but this idea was soon dismissed. Not everyone believed in the existence of molecules and atoms anyway. And if they did exist, they would be far too numerous and too small to produce the effect.

This made Einstein's explanation all the more remarkable. Yes, he said, it is collisions with air molecules that are causing the Brownian motion. But it is a statistical effect, not the result of individual collisions. On average, a smoke particle stays still because it is being bombarded by myriads of air molecules equally from all directions. But this bombardment is completely random. So over a period of a tenth of a second, say, more molecules may hit it from one direction than from another. As a result, instead of staying entirely motionless, the smoke particle is constantly being jostled about. What is more, Einstein said, you can actually use this Brownian motion to calculate roughly how many molecules there are in a given volume of air. All you need to know is the size of a smoke particle, and how fast it is jiggling about, and then you can calculate how many molecules must be hitting it each second. And from that you can work out how many molecules there must be in its immediate surroundings. This was not the only way one could estimate the number of molecules there must be, other methods were also being suggested at about the same time. But the Brownian motion method was the clincher, because it provided a figure much the same as that produced by other methods. Clearly this could not be just a coincidence, so molecules, and therefore the atoms that they are made of, must really exist.

A. Einstein, *Annalen der Physik*, 17, 549 (1905)

PHOTONS

Particles of light

Up until 1905, everyone thought they understood light. It consisted of waves in an 'electromagnetic field' that pervades the universe. Previously, this 'field' had been imagined as some sort of substance, referred to as the 'ether', but that had been shown to be wrong some twenty years previously. However, something rather strange had been noticed about these 'waves'. Round about the time that electrons were first discovered, in the late 1890s, it was found that if you shine ultraviolet light at a metal plate, it will knock electrons off it. This was called the 'photoelectric effect'. How could mere oscillations in an invisible field achieve this? This was the question to which Einstein applied his mind, and produced the paper that was to win him his Nobel Prize.

The cause of the photoelectric effect, Einstein said, lay in Planck's Quantum Theory. Every time a hot object gives off a quantum of energy in the form of light, that light itself takes the form of an indivisible packet—a particle to be called a 'photon'. The further up the spectrum the light is, the more energy each photon carries. Visible light, from red to violet, does not carry enough energy to produce the effect. In ultraviolet light however, a photon carries enough energy that when it hits an electron it can knock it right off its parent atom. The energy contained by a photon is given by the wonderfully simple equation

$E = h\nu$, where E is the energy, h is Planck's Constant and ν is the light's frequency or colour. To the world at large, this is not Einstein's most famous equation, but Einstein himself thought it was the best piece of work he ever did, and its implications for physics were scarcely smaller than the Relativity theories that were to follow.

The existence of photons was not proved by any single observation, but a whole series of experiments by the American Robert Milliken and also the Englishman Arthur Compton independently proved every detail of Einstein's hypothesis. Scientists have ever since had to live with the knowledge that light behaves both like waves and like particles, according to how you design your experiment. Stranger still, it is not just light that has this split personality. In the 1920s a young Frenchman called Victor De Broglie showed that every kind of particle, be it an electron, a proton, or even a whole atom, also behaves like a wave. This is the kind of discovery that leaves non-scientists baffled. It is easy to visualise a particle—just think of a tiny tennis-ball. And it is easy to visualise a wave. But how could anything be both a wave and a particle at the same time? The only answer that physicists have is to say, don't try to visualise these things—just check the results of your experiments against the theorists' equations, and see for yourself that it must be true.

A. Einstein, *Annalen der Physik*, 17, 132 (1905)

SPECIAL RELATIVITY

Why is the speed of light always the same? It just is. And $E = mc^2$

In 1905, astronomers had already known for some time that light behaves in a very strange way. It always travels in straight lines at a constant speed of 300,000 kilometres per second. Even if a star (for instance) is moving towards or away from the Earth at immense speeds, the light it throws off still reaches us at this invariable pace. This seemed impossible, but it had been proved by experiments and was undoubtedly true. Enter the young genius in the Zurich Patent Office, who addressed this problem in his usual way, with a paper as breath-taking as it was short (*see* Plate 3).

Einstein's brilliant solution to the conundrum had been to turn it on its head. Let's not ask why the speed of light is always constant, he suggested. Rather, let's assume that this is just one of nature's fundamental facts. Then let's see what we can deduce from that. What Einstein deduced was that if one applies a force to an object to accelerate it, one does not just make it go faster, one also makes it heavier. At the sort of speeds we are used to, the added mass is too small to detect. But as we approach the speed of light, less and less of the energy we expend goes to make the object move faster, and more and more of it goes to make the object heavier. Eventually, when an object has been accelerated to just below the speed of light, it cannot be made to go any faster at all, and any more energy expended on it simply adds to its mass. The conversion rate of energy into mass is given by the most famous equation in all of science, $E = mc^2$, where E is energy, m is mass, and c is the speed of light. c^2 is an enormous number, so this means that a huge expenditure of energy results in only a very small increase in mass. Or alternatively, a tiny loss of mass results in a huge release of energy—as when a nuclear bomb explodes.

It is often said that Einstein, one of the twentieth century's most ardent pacifists, had to bear responsibility for the atomic bomb. This is not really true. The bomb was a product of quantum physics rather than relativity theory. All Einstein did was sign a letter to President F. D. Roosevelt written by some other scientists, warning that the bomb was a possibility, and that Nazi Germany was believed to be working on it. Einstein signed the letter because he was the most famous physicist in the world at the time, and his signature was the best way to ensure that Roosevelt read the letter and took it seriously. Einstein agreed that America must build a bomb before Hitler did, but he was horrified when bombs were actually dropped on Japan, and after the Second World War he devoted much of the remaining ten years of his life to campaigning for nuclear disarmament.

Einstein, *Annalen der Physik*, 18, 639 (1905)

SUPERCONDUCTIVITY

Zero resistance to electric current at extreme low temperatures

Electrical resistance is a perennial problem in electrical and electronic engineering. Electric current consists of electrons passing through a substance, hopping from one atom to the next. With each hop, they lose a little energy in the form of heat. All substances, even the best conductors such as copper, give some resistance. However, when a way was found of abolishing resistance, it remained little more than a laboratory curiosity for over seventy years, and even now the difficulties in putting it to practical use seem immense.

Heike Kammelingh Onnes was a Dutch physicist, who pioneered the study of materials at extreme low temperatures. In 1908 he achieved the lowest temperatures yet by condensing helium gas to form a liquid, at 269 °C. This is only 4 °C above the theoretical (and impossible) 'absolute zero' of 273 °C, at which atoms would almost stop vibrating, and have no heat beyond a minimum 'zero point energy'. Onnes immersed various substances in liquid helium in order to study their properties under these extreme conditions. Among other experiments, he tried passing electric currents through various metals. His results came as a considerable surprise.

When testing mercury in this way, Onnes found that its electrical resistance abruptly disappeared completely at a temperature of 268.5 °C, or 4.5 °C above absolute zero. He soon found that several other metals also displayed this 'superconductivity' at extreme low temperatures. This phenomenon caused considerable interest, but research progressed slowly, and it was not until 1957 that an explanation was worked out by John Bardeen, Leon Cooper and Robert Schrieffer at the University of Illinois. They were duly awarded a Nobel Prize—which was Bardeen's second, his first one being for work on transistors some ten years earlier. They showed that in a superconductor, electrons form into pairs (called 'Cooper pairs'), which behave as single particles possessing the lowest possible amount of energy. Most everyday particles have a property called 'spin' but a Cooper pair does not. This means (among other things) that it can bounce past atoms without giving or taking any energy. The Cooper pairs link up, and move through the superconductor together, going straight past the atoms instead of hopping from one to the next.

Putting superconductivity to any practical use was clearly going to present problems. Liquid helium is still the only practical way to reach the very low temperatures needed, and it is expensive and difficult to handle. In the 1960s a few computers were built with superconducting circuitry to increase their efficiency, but the arrival of silicon chip technology made this idea largely obsolete. Since then, however, they have found a use in medicine: superconducting coils are used in the MRI (magnetic resonance imaging) scanner. This machine is mainly used for detecting tumours, especially in the brain. But it is only since the discovery of 'high-temperature' superconductors in 1986 that there has been any hope of superconductivity being used more widely, and by the end of the twentieth century this technology had still not progressed beyond the laboratory stage.

H. Kammelingh Onnes, *Communications—Physical Laboratory University of Leiden*, 120b (1911)

THE ATOMIC NUCLEUS

Rutherford's greatest discovery

In 1907 Rutherford left Mcgill University in Canada, to become Professor of Physics at Manchester University, which was already establishing itself as a centre of scientific excellence. There he set up a research team of graduate students, in the way that university laboratories have done ever since. This was the beginning of 'big science', and much of the history of twentieth century science is the story of the rise of the big research teams and the decline of the lone experimenter.

Rutherford had discovered that much of the radiation given off by radioactive elements consisted of 'alpha-particles'. He quickly proved that these were nothing less than positively-charged helium atoms, moving at an appreciable fraction of the speed of light. He found that they could pass through gases, and even thin barriers of solids. However, they did not quite move in straight lines. There was always a certain amount of 'scattering', as whenever they passed through any kind of barrier they would fan out like the pellets from a shot-gun. Rutherford set his assistant Hans Geiger (inventor of the Geiger counter) to work on this effect, together with a young graduate student called Ernest Marsden.

It was Marsden who made the original discovery. He was firing alpha particles at a sheet of gold leaf just 0.00006 millemetres thick. Most of the particles passed through, with some degree of scattering. But just a few of them bounced straight back. This was extraordinary. As Rutherford said, 'It was as though you had fired a fifteen-inch shell at a piece of tissue paper and it had bounced back and hit you'. How could such a thin barrier turn back a particle travelling so fast? And why did it only turn back about one in every eight thousand of them? Marsden and Geiger published their results in 1909, and eighteen months later Rutherford hit on the explanation. It was a true 'eureka moment'.

Until then, atoms had been thought of as being spheres, with the electrons embedded in them like raisins in a pudding. Rutherford now realised that this was wrong. Almost all of the bulk of an atom is in a tiny, positively-charged nucleus, with the rest of its volume made up of a cloud of orbiting electrons. Most of an atom is empty space. When alpha-particles travel through a solid, such as a sheet of gold leaf, almost all of them pass through the empty parts of the atoms, although the positive charges of the nuclei may deflect their path a little as they go past. But just occasionally, an alpha-particle will score a direct hit on a nucleus, and then it bounces back the way it came.

Rutherford sometimes admitted that he always envisaged atoms, nuclei, alpha-particles and electrons as being like miniature billiard balls. All his experiments had a direct, common-sense feel to them. Nonetheless, they confirmed much of the 'new physics' of Bohr and the quantum theorists, who had given up trying to visualise these particles at all.

E. Rutherford, *Philosophical Magazine*, 21, 669–688 (1911)

COSMIC RAYS

High-energy particles from outer space

When radioactivity was first discovered in the closing years of the nineteenth century, one of the first things noticed about it was that radiation carried an electrical charge. In those days, static electricity was still an exciting phenomenon, which could be studied with quite simple apparatus. One well-known device was the 'electroscope', still used in school physics laboratories until recently. It is simply a glass-sided box containing a piece of gold leaf attached to a conducting metal plate, that can be charged with static electricity or discharged again at will. When the plate is charged, the gold leaf moves away from it. It was noticed that if you charge an electroscope and then place it next to a piece of radioactive material, it soon loses its charge. However, an electroscope will lose charge slowly even if there is no source of radiation nearby. It was small indications like these that led a physicist called Charles Wilson to produce a radical suggestion: maybe there is radiation all around us, arriving from outer space. The quantity would be very small, but each ray would be extremely energetic, enabling it to penetrate the atmosphere and considerable amounts of solid matter.

This startling hypothesis was confirmed in 1912, by an Austrian physicist called Victor Hess. The first scientists to study radioactivity did many highly dangerous experiments, but usually they were quite unaware of the risks they were taking.

Nobody knew the effects that radiation could have on the human body. However, Hess took a risk of a different order. He did his experiments while flying in a balloon. He went up to eight kilometres in an open basket, exposing himself to extreme cold and lack of oxygen, as well as all the usual balloonist's problems of getting down again safely. Understandably, the experiments he performed on his flights were deliberately kept very simple. He took a number of electroscopes up with him, and observed how quickly they would lose an electric charge at different altitudes. Sure enough, the higher he went, the quicker they lost their charge. This could only mean that the loss of charge was being caused by something coming from outer space, that was gradually filtered out as it passed through the atmosphere.

We now know that most cosmic rays are protons, with some alpha-particles and also gamma-rays. However, even now we do not know where they originate. Some may come from within our galaxy, probably from exploding stars called supernovae. At least some seem to come from the very furthest reaches of space. They can be immensely energetic: when they hit molecules in the atmosphere, the result is showers of exotic sub-atomic particles. It was not until the late 1950s that machines were built powerful enough to produce these particles artificially, so their behaviour could be studied under controlled conditions. This research has led to theories about the basic nature of matter, including the quarks—the most fundamental particles of all.

V. Hess, *Physische Zeitung,* 13, 1984 (1912)

GENERAL RELATIVITY

What is gravity? Curved space and time

Ten years had passed since Einstein had rocked the scientific world with his three papers on Brownian motion, photons, and special relativity. He had left the Zurich Patent Office, and become a full-time mathematician. His reputation was huge, but surely even he would not achieve anything as revolutionary again. However, that is what he did, with a brand-new theory of gravity.

Newton's famous theory contains a problem right at its heart. It calls gravity a 'force' that objects exert on one another over a distance. But how can the force of gravity be transmitted across intervening space? Newton himself had been unhappy about it. Also, one small anomaly just did not fit the theory: the orbit of the planet Mercury did not quite follow the path Newton predicted for it.

Einstein applied his mind to the problem, and had a brainwave. 'There occurred to me the happiest thought of my life', he later wrote, *'because for an observer falling freely from the roof of a house there exists*—at least in his immediate surroundings—*no gravitational field'* (his italics). If you jump off the top of a house, you are completely weightless until you hit the ground below. So, said Einstein, gravity is not a force at all, it is a distortion of space and time. If you can do geometry in not just three dimensions but four (three of space and one of time), you can show that the planets orbiting round the Sun (for instance) are really travelling in straight lines! The Sun's mass has distorted the space and time in its vicinity, so that the planets' paths appear curved. It is like watching the shadow of an aeroplane flying over a hill. The shadow describes a curved path following the undulations in the ground, while the aeroplane is actually flying in a straight line. When Einstein did the calculations, he immediately found that his new theory not only explained everything that Newton's theory had explained, but it explained the orbit of the planet Mercury as well. Elated, he quickly published another of his famously short papers.

Not surprisingly, most people have trouble in swallowing General Relativity theory. It seems to go against all our intuition and experience. But it has repeatedly been proved to be true. Perhaps the most compelling proof of General Relativity must be the Global Positioning System. Anyone can now buy a GPS satellite receiver that will tell them their exact position on the Earth's surface to within a few metres, by receiving a signal from a satellite passing a few thousand kilometres overhead. For this to work, the satellite must transmit a signal giving its exact position, and also the exact time. But according to Einstein, the satellite, being further from the Earth's centre of gravity, travels through time very slightly slower than we do down here on the ground. So the clock that the satellite carries must be set to run slightly fast to compensate. Otherwise, our readings from our GPS receivers would be inaccurate by several metres.

A. Einstein, *Prussische Akademie der Wissenschaft*, p. 844 (1915)

THE PROTON

A chip off an atomic nucleus

Rutherford had shown in 1911 that an atom consists of a tiny massive nucleus, surrounded by a cloud of (almost) weightless electrons. The next question was: What is the nucleus made of? Different elements have nuclei of different weights, which suggests that the nucleus must be made of a number of smaller particles. Rutherford set his team at Manchester to work on this question. He had already worked out that the 'alpha-particles' given off during the radioactive decay of radium were in fact helium atoms minus their electrons, and he had found that if these were fired at a heavy element such as gold, they would bounce off. But what happened if alpha-particles were fired at lighter elements?

This research programme unfortunately coincided with the First World War. Rutherford went to work for the military, developing a listening apparatus for detecting submarines, but he returned to his laboratory at Manchester whenever he could. He set his colleague Ernest Marsden to work on experiments in which he exposed hydrogen gas to the fast-moving alpha-particles. Most of them passed straight through the gas, but among the alpha-particles emerging from the gas chamber were other, much smaller, particles travelling at very high speeds. In 1918 Rutherford's involvement with the military started to wind down, and he had time to investigate Marsden's particles further. He showed that they had a positive electric charge, and were in fact the nuclei of hydrogen atoms that had been hit full-square by the alpha-particles. He then tried the experiment again, only this time replacing the hydrogen gas with ordinary air. The result seemed extraordinary. Fast-moving hydrogen nuclei emerged from the gas chamber, just as if it contained hydrogen. But there is no hydrogen in ordinary air, so where were they coming from? Air consists mainly of nitrogen and oxygen, together with much smaller quantities of carbon dioxide and water vapour. Rutherford investigated further, and showed beyond doubt that the hydrogen nuclei were coming from the nitrogen atoms.

It would seem that the nucleus of a nitrogen atom consists at least in part of a number of hydrogen nuclei jammed together, and a direct hit from an alpha-particle can sometimes send one or more of these hydrogen nuclei flying off. The hydrogen nucleus itself must be one of the indivisible building-blocks from which larger nuclei are made. Rutherford decided that it needed a name of its own, and in 1920 he coined the name 'proton' for it.

Rutherford's experiment has been described as the first successful smashing of the atom. This is perhaps an exaggeration—'chipping the atom' would probably describe it more accurately. But such was the impression it made, that Rutherford was soon afterwards made Head of the Cavendish Laboratory at Cambridge, then the world's foremost physics laboratory. Until his death sixteen years later, he presided over a research programme that confirmed the models being developed by the quantum theorists, and gave us our basic knowledge of the atom's structure.

E. Rutherford, *Philosophical Magazine (6th series)*, 37, 581–587 (1919)

THE QUANTUM ATOM

Why are atoms stable? Because of quantum theory

Rutherford had shown that atoms consist mostly of empty space, with almost all of the mass concentrated in a tiny nucleus, surrounded by electrons. He envisaged the electrons orbiting the nucleus in the same way as the planets orbit around the Sun. This sounded very neat and logical, but it had a big problem. It had been known for over twenty years that when electrons oscillate back and forth, they give off energy in the form of electromagnetic waves. This is what is happening in the aerial of a radio transmitter. But in Rutherford's atom, the electrons are oscillating at immense frequencies, so in theory they should shed all their energy in a small fraction of a second and fall into the nucleus. This clearly is not happening, so Rutherford's model must be wrong somewhere.

It is at this point in the story that Rutherford's team at Manchester was joined by a man who was to become perhaps the most influential theorist of the twentieth century: Niels Bohr. He had been working at Cambridge on how hydrogen emits light at specific frequencies when it is heated, and was using quantum theory to find an explanation. He now extended this work, and showed that quantum theory gave the answer to Rutherford's problem. An electron cannot lose energy continuously and go spiralling down into the nucleus: it can only gain or lose energy in steps. Each step locks it into a particular stable orbit around the nucleus, and as long as it stays in that orbit it will not emit any radiation. In this way, atoms remain stable.

Bohr's discovery had implications reaching far beyond the confines of theoretical physics. In the early 1930s, the chemist Linus Pauling showed that this behaviour of electrons explained how atoms link together to form molecules, when one atom will fill an empty orbit by sharing an electron from another atom. The quantum nature of electrons also underlies our understanding of semiconductors—the crystals from which transistors are made.

After Bohr had worked with Rutherford for a few years, he returned to his native Denmark, and set up an Institute of Theoretical Physics in Copenhagen, largely financed by the Carlsberg Brewery. Most of the great theorists of the mid-twentieth century studied there at one time or another, in a pleasant and informal atmosphere. Bohr himself devoted his energies to developing his theory of 'complementarity', which holds that in all particle physics the methods one uses to observe the particles must be included in the equations one uses to describe them. When Denmark was overrun by the Nazis in the Second World War, Bohr stayed on for as long as he safely could, despite being an outspoken pacifist of Jewish ancestry. Eventually he had to be smuggled out aboard a British bomber in a hair-raising night-time snatch, just in time to avoid being arrested by the Gestapo. He spent most of the rest of the War in America, advising the physicists working on the atomic bomb project.

N. Bohr. *Physische Zeitung*, 24, 106 (1923)

THE UNCERTAINTY PRINCIPLE

The limit to what we can know

In the early seventeenth century, the philosopher René Descartes made the case for determinism. The universe, he said, is like a huge clockwork machine. Once it was set in motion, its entire future history was determined by the strict law of cause and effect. For the next three hundred years, this was the view of scientists and philosophers alike. Then in the twentieth century came quantum physics and determinism was finally abandoned. Central to this revolution was Heisenberg's Uncertainty Principle, which states that some things cannot possibly be known with complete accuracy.

Werner Heisenberg was a classic example of the young scientific genius. In 1926, at the age of just twenty-four, he produced a mathematical theory called Matrix Mechanics, which aimed to apply quantum theory to the behaviour of electrons. But within a year, a rival theory called Wave Mechanics was produced by Erwin Schrödinger, and for a moment Heisenberg's achievement seemed to be eclipsed. Then it was shown that the two theories were equivalent—they arrived at the same predictions, only by different methods. At this point, Heisenberg went to work with Niels Bohr at his Institute of Theoretical Physics in Copenhagen. After discussions with Schrödinger, Bohr had concluded that Wave Mechanics failed to predict completely the movements of electrons. There was more work to be done.

Through the winter of 1926–7, Heisenberg worked at producing a complete quantum mechanical theory to explain the movement of atoms and electrons, just as Newton's physics had explained the movement of more everyday objects. It proved impossible. One can only explain things that can be observed in an experiment, and no experiment can ever show the exact position and the exact velocity of an electron (or any other particle) simultaneously. This, wrote Heisenberg, is the Uncertainty Principle: *'The more precisely we determine the position, the more imprecise is the determination of velocity in this instant, and vice versa.'* This spelt the end of Descartes' clockwork universe. In his idea that if we know the present, we can calculate the future, 'it is not the conclusion that is wrong, but the premise'—we never can know the present exactly enough, so we can never predict the future. Determinism is dead.

Heisenberg's subsequent career was controversial. Many of Europe's leading quantum physicists were Jewish, and fled to Britain and America to escape from the Nazis. Heisenberg, as a Gentile, decided to stay on in Germany, despite being hounded by the SS, who considered quantum physics to be a Jewish perversion. He became involved in nuclear research, where his chief interest was in building a reactor. However, he well knew that his employers wanted a bomb. He made little progress. He believed that scientists should keep out of politics as much as possible, and thoughout the Nazi years he made it is his chief aim to keep physics going in Germany, even at the expense of collaborating with a regime that he found obnoxious. Many of his exiled colleagues found it hard to forgive him.

W. Heisenberg, *Zeitschrift fur Physik*, 43, 172–198 (1927)

THE POSITRON

The first particle to be predicted before it was seen

Some people think that science is just a matter of amassing facts. However, to scientists themselves the chief aim is to make theories to explain the facts. A scientific theory should make clear predictions that can be tested—it is not valued nearly so highly if it seems to explain everything we know already, but does not suggest what discoveries may await us in future. This was the position that physicists had reached in the 1920s. They were finding new sub-atomic particles, but their explanations were completely post-hoc. They needed a theory that would tell them what further particles were waiting to be discovered. One finally arrived in 1930.

The British theoretical physicist Paul Dirac was working on a way to unify Einstein's theory of Special Relativity with the Quantum Mechanics theory developed by Werner Heisenberg and Erwin Schrödinger. The two theories had been worked out separately, and at first sight their mathematics seemed incompatible. However, Dirac found that with a little adjustment Quantum Mechanics could be made to fit in with Special Relativity, and what was more, the new version could be tested. It clearly predicted that there should exist a particle which had never been seen before, a particle like the well-known electron, but its exact opposite. Where the electron carries a negative charge, this new particle should carry a positive charge. If this 'positive electron' were to collide with a normal electron, the two should annihilate each other, turning all their (tiny) mass into a flash of radiation in accordance with the famous equation $E = mc^2$.

So much for the theory—now the hunt was on to see if these 'positive electrons', or 'positrons' as they are now called, actually exist. They would surely be hard to find. There are so many normal electrons about, that a lone positron would be expected to last only a fraction of a second before it hit one and was destroyed. However, Dirac predicted that when cosmic rays hit the Earth's atmosphere, they should generate (among other things) electrons and positrons in pairs, flying in opposite directions. These were duly seen in 1931, by Carl David Anderson at the California Institute of Technology. However, he failed to spot that they were the particles of Dirac's theory. This was only realised the following year, when Patrick Blackett and Giuseppe Occhialini at Cambridge University repeated Anderson's experiments in more detail. Particle physics had come of age—it had moved on from listing particles as they were found, and was now predicting what would be found next.

We now know that not just electrons, but all other particles must have their corresponding opposites, or 'anti-particles'. When the universe first appeared in the 'big bang', in the first fraction of a second of time there was very nearly as much anti-matter as there was matter. However, the particles and anti-particles kept on colliding and annihilating each other, until there was no anti-matter left. Why there was very slightly more matter than anti-matter in the first place is something we still do not understand.

C. D. Anderson, *Science*, 76, 238–239 (1932)

THE NEUTRON

A chargeless particle with the mass of a proton

Today we all know that an atomic nucleus is composed of two kinds of particles: protons, which carry a positive electrical charge, and neutrons, which carry no charge. Each chemical element contains a specific number of protons, but the number of neutrons can sometimes vary slightly. Typically, a nucleus contains roughly as many neutrons as protons. If it contains too few or too many neutrons, it becomes unstable and radioactive.

Arriving at this picture was the great achievement of the nuclear physicists of the early twentieth century, who devised different ways of knocking fragments off atomic nuclei. Their first method was simply to study the radiation from radioactive elements; then they used this radiation to bombard atoms, and see what happened. The first sub-atomic particle to be described by these means was the positively-charged alpha-particle, which we now know consists of two protons and two neutrons. Next to be seen was the proton, first observed by Rutherford in 1919. However, throughout the 1920s, physicists increasingly came to believe that there must be parts of a nucleus that did not carry a positive charge: their calculations strongly suggested that there must be neutral particles as well. The problem was in devising a means to prove their existence. Rutherford described the neutron as 'like an invisible man passing through Piccadilly Circus: his path can only be traced by the people he has pushed aside'. He gave the problem to his colleague James Chadwick, who spent several years trying various ways to make a neutron show its presence. He later described some of his experiments as being 'so far-fetched as to belong to the days of alchemy'. Success finally came in 1932.

Chadwick was not the first person to see neutrons: that distinction goes to two Germans named Walter Bothe and Herbert Becker. In 1930, they found that if the light element beryllium was bombarded with alpha-particles, a new radiation with no electric charge was given off. Chadwick, suspecting the true nature of this radiation, performed a series of experiments and proved that it consisted of particles of the same mass as protons, but without their charge. The search was over.

For most practical purposes, the description of the atom was now substantially complete. The composition of the nucleus—protons and neutrons—had been found by Chadwick and Rutherford, while Niels Bohr had worked out the behaviour of the surrounding electrons. Further research on even smaller-scale structure had to wait until after the Second World War, culminating in the detection of quarks in the 1970s. (Whether this knowledge will ever have any practical application remains to be seen, but if so it probably lies far in the future.) Meanwhile, the neutron was the key to the nuclear chain-reaction: in 1938 it was found that one neutron could split a uranium nucleus, liberating two more neutrons that could repeat the process. This is the principle behind nuclear energy as used in both reactors and bombs, and research into it was to occupy most physicists throughout the 1940s.

J. Chadwick, *Nature*, 129, 312 (1932)

NUCLEAR FUSION

Rutherford's last triumph

By the 1930s, Rutherford was the Grand Old Man of physics. The farmer's son from New Zealand was now Lord Rutherford of Nelson, President of the Royal Society, head of the Cavendish Laboratory at Cambridge, and with a Nobel Prize under his belt (*see* Plate 2). With his commanding presence and booming antipodean voice, he was as well-known in the Athenaeum Club as in the Senior Common Rooms of Cambridge. There were many affectionate anecdotes about him. Scientists at that stage in their careers often become administrators, their best ideas behind them, leaving the actual research to their younger colleagues. Rutherford was indeed heavily involved in the politics of science, promoting the cause of nuclear physics in the corridors of power. However, he found time to carry on with his experiments until the end of his life, and in 1933 he produced one most of his stunning achievements: nothing less than a bench-top demonstration of the H-bomb.

The previous year, Rutherford's pupils Cockroft and Walton had split the atom, using a generator capable of accelerating protons and other positively charged particles up to hitherto unheard-of speeds. Rutherford's next move was to mount a team project, using the new machine to fire positively-charged atoms of 'heavy hydrogen'—deuterium—at a thin sheet of lithium. The result was nuclear fusion, of precisely the kind that was to be used in the bomb. Whenever a deuterium nucleus hit a lithium nucleus head-on, the two momentarily merged before splitting into two helium atoms moving in opposite directions at immense speeds. The following year, the team repeated the experiment, this time fusing two deuterium atoms together. As Rutherford wrote: 'an enormous emission of fast protons was detectable even at energies of 20,000 volts. At 100,000 volts the effects are too large to be followed by our amplifier. . . . It seems probable that the (deuterium ions) unite to form a new helium nucleus . . .'

One might imagine that Rutherford would have realised that people would soon be seeking applications for the new sources of energy he had discovered, but in fact he had no inkling of it. This may have been because he did all his experiments with very small-scale apparatus, often held together with sealing-wax, built on his laboratory bench. So in an address to the British Association in 1933, he said: 'These transformations of the atom are of extraordinary interest to scientists, but we cannot control atomic energy to an extent which would be of any value commercially, and I believe we are not likely ever to be able to do so. A lot of nonsense has been talked about transmutation. Our interest in the matter is purely scientific.' Rutherford was not trying to find new sources of energy. He was simply trying to find out what atoms were made of, by smashing them to pieces and weighing the fragments. He was never to see atomic energy put to any use, because he died suddenly in 1937, at the age of sixty-six.

Rutherford et al, *Proceedings of the Royal Society A*, 141, 722–733 (1933)

NUCLEAR FISSION

The beginning of the Nuclear Age

The answer to the question 'Who was the first person to split the atom?' depends on how one defines 'splitting the atom'. In 1919 Ernest Rutherford used alpha-particles (fast-moving helium nuclei) to knock individual protons off nitrogen atoms, but that was not so much splitting the atom as chipping it. In 1932, John Cockroft and Ernest Walton, working in Rutherford's laboratory in Cambridge, fired high-speed protons at lithium, and saw the lithium nuclei break up into alpha-particles. The novelist C. P. Snow later described Cockroft as 'skimming down King's Parade, and saying to anyone whose face he recognised, "We've split the atom! We've split the atom!" '. However, Cockroft himself said that this story 'owed more to fancy than fact'.

True nuclear fission, as used in reactors and bombs, was first achieved in 1938 by Otto Hahn and Fritz Strassmann, working at the Kaiser Wilhelm Institute for Chemistry in Berlin. They fired neutrons at uranium (the heaviest element to occur in nature), and found that a direct hit on the nucleus would cause it to break into two almost equal halves. This was a big surprise, because they could not understand how a single neutron could make a strong enough impact. The explanation was worked out by Lise Meitner, a former colleague of Hahn's. (Being Jewish, she had fled her native Austria some months earlier when Hitler marched in, and was now living in Sweden.) She calculated that the uranium nucleus must behave not like a solid lump, but rather like a liquid drop that shuddered and split when hit by a neutron. The two halves, both having positive electric charges, would then repel each other and fly apart at great speed. Most intriguing of all, however, was the possibility that when the nucleus split, there would be one or more neutrons left over, which would then hit other nuclei and cause them to split in turn. This 'chain reaction' was soon proved to occur.

The possibilities opened up by this discovery were not lost on the scientists involved. It meant that a single gramme of uranium could release energy equivalent to burning three tonnes of coal, and this could become a runaway process resulting in a huge explosion. Among the many Jewish scientists now living in America, Britain and Scandinavia, there was alarm that this research was being done in Nazi Germany. The Hungarian Leo Szilard, who had first worked out the theory that Hahn and Strassmann were testing when they made their discovery, drafted a letter to President Roosevelt, warning him of the dangers. In order to ensure that Roosevelt read the letter, he persuaded the world-famous Professor Einstein to sign it. This letter eventually led to the Manhattan Project and the building of the first nuclear reactors and bombs. The Nazis, meanwhile, never concentrated their researches in one big project, and as a result never got anywhere near to producing a bomb.

O. Hahn and F. Strassmann, *Naturwissenschaften*, 27, 111 (1939)

THE TRANSISTOR

Quantum effects harnessed to cause a social revolution

People sometimes ask: What is the point of all this science? It is easy to justify research carried out with a clear goal, as in medicine. But how can physicists justify the time and expense of their research, which consists of accumulating apparently useless knowledge for its own sake, often at great expense? People who ask such questions might ponder the story of the transistor, a device that has changed the world arguably as much as any other scientific development of the twentieth century. In addition to its use in radios, televisions and hi-fi sets, it has become ubiquitous in its miniaturised form on the silicon chip. The very way in which we take it for granted testifies to its place in the infrastructure of Western civilisation. And yet, the transistor effect was first predicted by the theoretical physicists Enrico Fermi and Paul Dirac, who in turn based their theory on Niels Bohr's quantum mechanical theory of electrons. Bohr was the purest of academics, and his Institute of Theoretical Physics in Copenhagen was most people's idea of the perfect ivory tower. Neither he nor the Carlsberg Brewery, which financed the Institute, had any idea that practical applications might come out of his theorising.

The possibility of the transistor was first realised by William Shockley at the Bell Telephone Laboratories in New Jersey, USA. An entry in his notebook dated Christmas Eve,

1939, begins with the words: 'It has occurred to me that an amplifier using semiconductors rather than valves is in principle possible . . .' 'Semiconductors' are crystalline substances such as germanium or silicon, that have properties intermediate between metals and non-metals. In Bohr's model of electrons filling discrete orbitals in each atom, the atoms of semiconductors have just a few empty orbitals. When an electric current flows through a semiconductor, electrons hop from one atom to the next, leaving empty orbitals behind them. According to the Fermi-Dirac theory, these empty orbitals, called 'holes', behave like particles travelling in the opposite direction to the electrons. The electric current behaves simultaneously like a flow of electrons in one direction and a flow of holes in the other. The trick of a transistor is to use the flow of holes between two contact points to influence the flow of electrons from a third contact point. A weak signal, consisting of slight fluctuations in the flow of holes, becomes a strong signal in the form of big fluctuations in the flow of electrons. Shockley was not the first to realise the nature of current in a semiconductor, but he was the first to see its full potential.

After the Second World War, Shockley formed a team with John Bardeen and Walter Brattain to put his idea into practice. After a few false starts Bardeen and Brattain built a germanium transistor, filed a patent application and published a paper in *The Physical Review*. Shockley, who believed that he should have been given chief credit for the idea, had his name on papers and patents on related research. All three were awarded the Nobel Prize in 1956.

J. Bardeen and W. H. Brittain, *Physical Review*, 74, 230–23 (1948)

THE NEUTRINO

All spin and no substance

Of all the strange quests that scientists set themselves, few can be as ambitious as the search for the neutrino: a particle with no mass and no electric charge, which can pass straight through the Earth as if it were not there.

The neutrino was first suggested in 1933 by Wolfgang Pauli, one of the great theoretical physicists of the first half of the twentieth century, whose particular interest was radioactivity. When a radioactive atom decays, parts of its nucleus break up. One kind of breakage is called the 'beta-decay', when a neutron breaks into two parts: a proton and a fast-moving electron, known as a 'beta-particle'. However, in this process a significant fraction of the energy involved seems to go missing. So Pauli suggested that it was being taken away in the form of an almost undetectable particle, whose only property was 'spin'. All sub-atomic particles behave as if they are rotating at immense speeds. Pauli's particle effectively consisted of nothing at all, spinning rapidly. Even his colleagues, well used to the seemingly preposterous nature of the sub-atomic world, found this hard to swallow at first.

Pauli was no experimenter. Indeed, his friends said he only had to walk into a laboratory for equipment to break down. So it was not until twenty years later that Frederick Reines and Clyde Cowan at the Los Alamos National Laboratory in Nevada, USA, hit on a way to make these ghostly particles show their presence. In a particularly neat logical sleight-of-hand, they did Pauli's mathematics in reverse, to see what would happen if a neutrino collided with a proton head-on—a very rare occurrence, but it should happen occasionally. Armed with the results of their calculations, they devised a special neutrino detector, consisting of several chambers full of water with cadmium chloride dissolved in it, and placed it where they might expect to find plenty of neutrinos—next to a nuclear reactor. Their calculations told them to expect to see tiny flashes of gamma-rays, occurring in pairs. These flashes duly appeared, and what was more important, their frequency depended on the power at which the reactor was running. The detector was shielded from all other kinds of radiation, so these flashes could only be caused by neutrinos.

Since this original experiment, several neutrino detectors have been built, mostly for catching neutrinos coming from outer space. Some come from stars, while others are left over from the 'Big Bang' at the beginning of time. Paradoxically, the detectors have to be built deep underground, to shield them from other kinds of radiation. The chief question asked about neutrinos is: do they really have no mass at all? At the close of the twentieth century, the answer is still not conclusively established, but the best guess is that they do indeed have a minuscule mass. This has major implications for cosmology. Neutrinos are so abundant, that if they weigh anything at all they must make up a substantial fraction of the entire mass of the universe.

F. Reines and C. L. Cowan, *Nature*, 178, 446–449 (1953)

THE HUNTING OF THE QUARKS

The ultimate building blocks of matter

At the beginning of the twentieth century, most people thought that atoms were the smallest particles of all. Then it was discovered that they were made of electrons, protons, and neutrons. Finally in the 1970s it was shown that protons and neutrons themselves are made of smaller particles still.

By the 1950s, many different kinds of sub-atomic particles were known. The familiar proton and the neutron had been joined by no less than six exotic siblings (all eight were called the 'baryons'), and also seven much lighter particles called 'mesons'. Not all of them are confined to the laboratory, either. Some occur in nature, when a cosmic ray from outer space ploughs into the Earth's atmosphere, leaving smashed atoms in its wake. A theory was needed to explain them all.

The solution was worked out independently by two physicists, Murray Gell-Mann and George Zweig. They turned to an abstruse branch of mathematics called 'group theory' (which has nothing to do with 'groups' as most people would use the word), and found that it could produce a very elegant model. It was as if all of these sub-atomic particles were made of two or three even smaller particles still. Gell-Mann, who has a whimsical imagination, called them 'quarks', for no particular reason. According to the theory, there were three different types of quark, called (again by Gell-Mann) Up, Down, and Strange.

(The theory has since been developed to include three more, called Top, Bottom, and Charm.). There is also an 'anti-quark' corresponding to each quark. Baryons are made of quarks, and mesons are made of one quark and its corresponding anti-quark each. A single quark can never exist on its own—it would be like having a piece of string with only one end. In this theory, our familiar proton is portrayed as three quarks—two Ups and a Down, and a neutron is two Down quarks and an Up.

The theory quickly met with success. It neatly predicted the existence of all the particles that had been discovered, and one more named the Omega Minus, which was duly seen in 1964. However, this only proved that the theory was a useful one. It did not prove that quarks really exist. Gell-Mann, for one, believed that they were just mathematical abstractions, which only existed on paper. Others, however, had their suspicions.

Jerome Friedman, Henry Kendall and Richard Taylor are the three names associated with the massive 'deep elastic scattering' experiment conducted in 1967, using the latest accelerator at Stanford University in the USA to fire electrons at protons with such force that they penetrated right inside. The results were clear-cut. Just as fifty-six years earlier Rutherford had found that deep inside the atom was a tiny hard nucleus, later found to consist of protons and neutrons, so Friedman, Kendall and Taylor found that inside a proton were three tiny hard bits. So we can now say with certainty that quarks really exist. And we are pretty sure that there is nothing smaller still—they are the ultimate building-blocks of matter.

J. Friedman et al, *Annual Review of Nuclear Science*, 22, 203 (1972)

THE W AND Z PARTICLES

Proof of the Standard Model of atomic forces

One of the crowning achievements of twentieth-century physics, after proving that atoms exist and finding the particles of which they are made, was to work out how they are held together; and also what forces are involved when atoms of radioactive elements like uranium decay and fly apart. The picture developed slowly, starting in the 1920s, but by the 1970s physicists were talking of a 'Standard Model' of the sub-atomic particles and the forces that govern their behaviour. Final vindication of the Standard Model came in the early 1980s, when two highly exotic particles, the W and the Z, were observed.

According to the Standard Model, there are three forces at work inside an atom: the Electromagnetic force, which binds electrons, protons and neutrons together; the Strong Atomic Force, which holds quarks together to form protons and neutrons; and the Weak Atomic Force, which is involved in certain forms of radioactive decay. Each force is carried by a particle. The Electromagnetic force is carried by photons—the particles that light and radio waves are made of. One can think of photons shuttling back and forth inside atoms, tying the electrons, protons and neutrons together. This was worked out by Enrico Fermi in the 1920s, in a theory known as Quantum Electrodynamics. Similarly, the Strong force is carried by particles appropriately named 'gluons'—these were first observed in 1979.

The Weak force, however, is much more enigmatic. According to the Standard Model, it does not hold anything together, and only comes into play momentarily on those occasions when one kind of particle decays into another. It can be carried by two kinds of particle, called the W and the Z. However, under normal circumstances these are 'virtual particles'—mathematical abstractions that only exist on paper. Only if real particles collide with each other at extremely high energies should real Ws and Zs momentarily appear. The Standard Model predicted that when this happened, they should be extremely heavy—a W particle should weigh as much as an entire atom of iron, and a Z would be even heavier. This mass is generated out of the energy of the collision, as described by Special Relativity theory.

It was the achievement of a team at CERN, led by Carlo Rubbia and Simon van der Meer, to detect W and Z particles in late 1982 and 1983. To do this, they used the Proton-Antiproton Collider, a giant machine constructed underground near Geneva, Switzerland, that generated beams of protons and antiprotons travelling round a circular track in opposite directions at close to the speed of light. Wherever a proton and an antiproton collided, they annihilated each other. Just occasionally the collisions created Ws and Zs, which then instantly decayed into less exotic particles. It was these decay products, flying out from the collision points at sharp angles, which gave clear indication of the existence of Ws and Zs, and thus confirmed a theory that had taken over sixty years to build.

G. Arnison et al, *Physics Letters B*, 122, 103–116 (1983); *Physics Letters B*, 126, 398 (1983)

BOSE-EINSTEIN CONDENSATE

Individual atoms merge their identities at extreme low temperature

At extreme low temperatures, matter behaves in very strange ways. The coldest possible temperature, known as 'absolute zero', is 273 °C. All atoms vibrate and move around, but as they get colder their vibration gets less and less, and at absolute zero they would, in principle, have no movement at all beyond a residual 'zero point energy'. However, it is never possible to reach the exact absolute zero: the nearest one can get is to within a few billionths of a degree of it. One phenomenon that can be made to occur at these temperatures is the Bose-Einstein condensate: a lump of matter containing thousands of atoms, that behaves as if it consists of just one atom of immense size.

As its name suggests, the Bose-Einstein condensate has a distinguished background. The underlying theory was first worked out by Einstein in 1924, who based it on ideas originally suggested by the Indian physicist Satyendra Nath Bose. Einstein noted that all atoms (and indeed all sub-atomic particles) can be described by the same mathematical equations as one may use to describe a wave. As an atom gets cooler, it can be thought of as increasing its wavelength. At extreme low temperatures, two or more adjacent atoms may have such long wavelengths that they overlap and interfere with each other, and eventually the atoms' waves could be made to merge together. If this happened, all the atoms would have exactly identical properties and behaviour, and could no longer be told apart. They would do everything in perfect synchrony, like a perfectly-trained dance troupe or soldiers on a parade-ground.

Most physicists viewed the Bose-Einstein condensate as an intriguing piece of theory, but did not expect that anyone would ever make one out of a group of atoms in a vacuum. However, in 1995 a team led by Eric Cornell and Carl Wieman at the University of Colorado did exactly that, followed shortly afterwards by Wolfgang Ketterle at the Massachusetts Institute of Technology, who had been working on the problem independently. To achieve this, they had to cool down some atoms of the element rubidium to a temperature colder than had ever been achieved before—just 20 billionths of a degree above absolute zero. This was done in a two-stage process, firstly slowing down the atoms' movement by bathing them in laser light of precisely the right frequency, and then confining them in magnetic fields designed so that the least cold atoms would leak out, leaving only the very coldest atoms behind.

Nobody knows yet where research on Bose-Einstein condensates may lead, but there are some exciting possibilities. In 2001, physicists at the University of Munich created a condensate on the surface of a silicon chip. This achievement brings nearer the much-dreamt-of 'quantum computer', whose central processor would exploit the quantum properties of a single atom, to do more than one calculation simultaneously. A Bose-Einstein condensate behaves like a single atom, but being thousands of times bigger, it could be much easier to 'wire it up' as part of an electronic circuit.

M. H. Anderson et al, *Science*, 269, 198–201 (1995)

SCIENCE IN THE YEAR 2000

This book was largely written during the year 2001, less than two years after the twentieth century ended. Inevitably, this caused problems in choosing which scientific discoveries to include. Some discoveries are immediately hailed as major breakthroughs, but in many cases it can take years or decades for a discovery's true importance to become apparent. Also, it sometimes happens that a discovery which was considered significant when it was first announced later turns out to be very much less so. For reasons such as these, it was hard to do justice to the scientific advances that were being made in the century's closing years. The amount of effort being expended on scientific research worldwide continues to grow year after year, so one might expect that each decade would see more major discoveries than the one before it. But out of the one hundred discoveries described in this book, only seven date from the 1990s. There was no lack of candidate discoveries from this period, but for many of them it is just too early to pronounce on their importance.

This closing section aims to give a taste of what is happening in science at the start of the twenty-first century, by looking at a handful of fields where especially promising advances are being made, or are anticipated in the next few years. Inevitably it is highly partial and subjective. Most scientists will doubtless say that their own field is entering a particularly exciting period at this time, and many of them will probably be justified in thinking so. A comprehensive overview of all their claims would be impossible. Here, instead, are just a few of the research areas that were making headlines in the year 2000.

Life Sciences

In 2000, all eyes were on the Human Genome Project. Techniques for reading off the coded information in DNA, first pioneered by Sir Frederick Sanger in the 1970s, had become automated to the point where it was feasible to download the sequence of the entire genome of a human being in only a few years. In the 1990s two projects to do this were running in tandem: the 'official' Human Genome Project, run by the US National Institutes of Health under the leadership of James Watson; and a commercial project run by Celera Genomics Inc, led by Dr. Craig Venter. Celera started work several years after the 'official' project, but both produced their initial 'rough drafts' of the human genome in the summer of 2000. Unsurprisingly, Celera's draft appears to be the rougher of the two. The National Institutes of Health have deposited all the information from their project in the online databank GenBank, where it can be read by anyone. Celera also announced that they would make their version freely available, but their first intention was to recoup the cost of the project by selling genetic information to the pharmaceutical industry. The human genome is itself not a discovery, merely a mass of information, but in years to come biologists and medical researchers expect to make many discoveries by trawling through it. One early finding is that we seem to have very much fewer genes than was previously estimated. Some studies suggest that the total may be no more than 35,000 separate genes, although others point to about twice that figure.

Homo sapiens is not the only living organism to have had its genome sequenced in the year 2000. The entire genomes of several other plants and animals, including the famous fruit fly *Drosophila melanogaster*, were also sequenced in that year. Genomes are of much greater use to biologists if they have several of them, because it then becomes possible to compare and contrast different genomes. We will certainly find out much about human genes by comparing them with known genes in other organisms. Fruit flies do not have nearly as many genes as humans do, but a high proportion of their genes are also present in humans, where they may not always perform exactly the same function. We even share a considerable number of our genes with plants, where the functions they perform may be different yet again. Studying how similar genes perform similar or different functions in different organisms will teach us much about how they interact with each other.

Elsewhere in biology, scientific journals in the year 2000 were full of references to 'apoptosis'. This is the name given to the process by which cells in a living organism are killed to order. It may seem strange that cells are actively programmed to die, rather than being left to survive as long as possible, but this is necessary for controlled growth and for maintenance of healthy tissues. In a human body, some 10 billion cells will die and be replaced each day. Some of these may have been invaded by bacteria or viruses, but many more will be perfectly healthy. In particular both the human nervous system and the immune system produce many more cells than they need, and then kill off a large proportion of them . Knowledge of the mechanisms that control this process could have many applications in medicine, for instance in treating diseases of the nervous system, or destroying cancers by making them 'commit suicide'. Research in this field started to make rapid progress in the last two years of the century. A particularly notable discovery was 'CD95 death receptors', molecular structures embedded in the cell's surrounding membrane that act as receivers to pick up chemical signals ordering it to commit suicide.

History of Life

The rate of progress in palaeontology was accelerating as the twentieth century closed. To the general public, perhaps the most exciting discoveries were more and more fossils of dinosaurs. The familiar ones, such as *Tyrannosaurus rex*, were all discovered in Europe and North America in the nineteenth century, with a further flurry of discoveries in Mongolia in the 1920s. The fossils from these regions are now well known, so in the 1980s and 1990s fossil hunters started searching elsewhere, and were soon turning up spectacular finds, notably in Argentina, China, and West Africa. Some of these were bigger than any previously known. Sauropods (long-necked plant-eating dinosaurs) from South America may have rivalled the biggest whales in bulk. In China, meanwhile, beautifully-preserved fossils of small carnivorous dinosaurs confirmed what had been widely suspected: many of them had feathers or something very similar. The feathers on *Archaeopteryx*, the famous 'missing link' between reptiles and birds, did not evolve suddenly from nowhere. Rather, they were an enlarged version of something that many reptiles had already.

In Africa, meanwhile, the search for humankind's recent ancestors was also making exciting progress. A most exciting find was the skeleton of an ape named *Ardipithecus ramidus*, discovered in 1993 near Afar in Ethiopia by Yohannes Haile-Selassie, a member of a team led by the veteran fossil-hunter Tim White. *Ardipithecus* is at least half a million years older than the famous 'Lucy' (who was found only a few kilometres away), and according to people who have seen it, it appears to be much more like a chimpanzee. However, the fossil is extremely fragile, and at the time of writing White and his colleagues are still piecing it together. It is hoped that it will throw light on the still-mysterious period when our own ancestral lineage diverged from that of the chimpanzees.

Earth and Environmental Sciences

With environmental concerns becoming a major political issue, the Earth sciences have a higher profile than ever before. By 2000, most scientists were convinced that the 'greenhouse effect' was genuine (although there were still a few doubters, not all of them with links to the oil and coal industries). Successively refined computer simulations all showed that the burning of fossil fuels was bound to increase the amount of carbon dioxide in the atmosphere, and that this would on the whole cause temperatures to rise all over the world. However, these models were also showing that the process was unpredictable. Global warming would lead to rapid shifts in weather patterns, as major ocean currents such as the Gulf Stream suddenly change course or even stop completely. Such shifts seem to have happened in the past. Studies of samples taken from the Greenland and Antarctic ice caps showed how rapidly temperatures had swung during the several Ice Ages of the past million years. It seems that these had started and ended quite abruptly. Nobody yet knows for sure what caused them, but it seems likely that some initially small perturbation in the Earth's weather systems had immensely magnified consequences. The implications are not reassuring for anyone trying to predict future changes. It seems that the global climate may be only partially stable, and it can 'flip' from one fairly-stable state to another quite unpredictably.

Even more extreme than the Ice Ages of recent times were the 'Snowball Earth' periods. In 1999 some geologists presented evidence that on several occasions in the distant past the Earth was covered in ice sheets that reached as far as the tropics, and might even have covered the entire planet. These reflected so much of the sunlight that normally warms the Earth, that the planet became locked into prolonged periods of intense cold. Eventually, each 'snowball' episode ended when a period of volcanic eruptions spewed large quantities of carbon dioxide and other 'greenhouse gases' into the atmosphere, causing the ice sheets to melt as suddenly as they had arrived. The most recent 'snowball' episode is said to have ended some 570 million years ago, just before fossils started to become abundant. The Snowball Earth scenario has been debated intensively, and it should be said that the evidence for it is not yet conclusive. Climatologists have devised computer simulations, and these all seem to show that it would be unlikely, if not impossible, for the Earth to freeze over entirely. The tropics at least would have remained free of ice.

Chemistry

The areas of most interest, to non-chemists at least, remained fullerenes and nanotubes (*see* 'Buckminsterfullerene', page 196), and high-temperature superconductors (*see* page 198). In 2000 both seemed on the verge of having major practical and commercial applications. There were even some fullerene compounds that had been found to have superconducting properties. Unfortunately, they did not superconduct at temperatures high enough to be industrially useful, but they had the advantage of being much more pliable than the higher-temperature superconductors, most of which are very brittle and difficult to work with.

One particularly exciting discovery made in 2000 was a new way of extracting the metal titanium from its ore. Titanium is a highly abundant element, with properties that make it very attractive to engineers. It is almost as light as aluminium, but much stiffer. It also has a very high melting point, and hardly corrodes at all. In the second half of the twentieth century it was used extensively in building supersonic aircraft, but it was too expensive to be used for more everyday purposes—it cost about six times as much as stainless steel, for instance. (Its oxide, however, is familiar as the pigment used in most brands of white paint.) The problem with titanium has been extracting it from its ore. The method used until now has been the Kroll Process, which involves converting titanium ore into titanium tetrachloride, and then reacting this with molten magnesium to make metallic titanium and magnesium chloride—not a simple process at all. But in 2000, a team led by Derek Fray at Cambridge University quite accidentally discovered another method, that could cut the cost of producing titanium by as much as three-quarters. They were experimenting with electrolysis, as a possible way of removing the thin film of oxide that forms on the surface of any piece of titanium. They passed a current through an electrode coated with titanium oxide, immersed in a bath of molten calcium chloride. To their surprise, the titanium oxide quickly shed its oxygen and became pure metallic titanium. If this process can be made to work on an industrial scale, we may expect titanium to become as universal and familiar in the twenty-first century as aluminium was in the twentieth century, and cast-iron was in the nineteenth.

Astronomy

Astronomy is going through an exciting time, largely as a result of greatly improved telescopes, often orbiting in space. The Hubble Space Telescope is the most famous of these, but other instruments have also been sent into orbit to record infrared, ultraviolet and even gamma-rays. The Hubble Telescope itself is also becoming rivalled by telescopes on the ground equipped with 'adaptive optics', which enable them constantly to readjust their focus to compensate for turbulence in the atmosphere. By this means, they can obtain pictures almost as clear as those taken by a telescope in outer space. Meanwhile, the development of 'interferometry'—methods of pooling data recorded from two or more widely-separated telescopes—has enabled astronomers to see distant objects in much finer detail, to the point where they can very nearly see planets orbiting around distant stars.

On the large scale of cosmology, two questions stand out: these are referred to as the 'cosmological constant' and the 'missing mass'.

The 'cosmological constant', if it exists, is a factor that has to be included in the equations of General Relativity theory to describe a force that makes the universe expand. It has been known ever since Hubble's discoveries in the 1920s that the universe is getting bigger. However, ever since the Big Bang theory was vindicated in the 1960s it has been assumed that this expansion is gradually slowing down, under the force of gravity. But in the 1990s observations were made of some extremely distant (and therefore ancient) quasars, which seemed to show that the expansion is actually accelerating. This would seem to imply that there is some kind of 'anti-gravity' at work, actively pushing the universe apart. Further observations are still needed to confirm this, but if it is true, then a whole new branch of cosmological theory may have to be developed to explain it.

The 'missing mass' seems an even more intractable problem. Studies of galaxies' gravitational fields show that they are about ten times heavier than the total mass of all the stars, planets, black holes, dust, gas, and so on that they contain. What gives them all this extra mass? Cosmologists are generally forced to conclude that there are some kinds of structure in space that are not made of conventional matter, but which nonetheless have strong gravitational fields. Various theories have been devised, often based around the idea that there are in fact more than four dimensions of space and time. However, it has proved very difficult to frame these theories in ways that can be tested at all.

Nearer to home, the year 2000 saw increased interest in the possibility of extraterrestrial life. Although there is still no sign of intelligent, or even particularly complex, life forms in outer space, many scientists think it possible that simple organisms may occur on other planets in our Solar System. Two findings have triggered this renewed interest. The first was evidence that Europa, one of the moons of the planet Jupiter, has liquid water on it. This moon is totally covered with ice, but close-up photographs taken by space probes strongly suggest that this ice may be floating on an ocean of liquid water. This in turn would imply that something is keeping Europa's interior warm, and where there is both heat and water, life could conceivably flourish.

The other evidence for extraterrestrial life was claimed to come from a Martian meteorite. It has been known for some time that a few of the small meteorites that occasionally fall to Earth are in fact fragments of rock from Mars, which were blasted off that planet millions of years ago by the impact of an asteroid. One of these, found in 1984 embedded in ice in Antarctica, was said by some to contain fossil remains of bacteria-like organisms. Most of this supposed evidence has not withstood closer scrutiny, but in 2000 it was widely agreed that the case was still not finally closed. Photographs of Mars taken by space probes strongly suggest that there were once large amounts of liquid water on it, and there is also evidence that large quantities of ice may exist beneath its sandy surface. So although Mars today is inhospitable to life, being too cold and having insufficient atmosphere, it could possibly have supported life in the past. If there ever were life on Mars, it probably originally came from Earth on board a meteorite, or possibly both Earth and Mars were seeded with life from outer space—a scenario once considered absurd, but which is now getting increasing support from mainstream scientists.

Physics

In 2000, the Standard Model explaining sub-atomic particles and the ways in which they interact was considered to be fully vindicated. So many of the predictions it made had been proved right that no-one saw any need to change it or replace it. One major prediction, however, awaited confirmation: the existence of an all-pervading 'Higgs Field', which according to the Standard Model is what gives all the different particles their masses. The Higgs Field should show its presence in the form of a very massive 'Higgs Particle', a distortion in the Field that should appear when two more ordinary particles collide with sufficient force. The world's biggest particle collider, the LEP (*see* Plate 5) near Geneva, Switzerland, was believed not to be quite powerful enough to generate Higgs Particles, but in the autumn of 2000 some researchers analysed data from it and claimed to see evidence of them in small numbers. The LEP was due to be switched off at the end of 2000, so that an even more powerful machine could be built in its place. However, the quest for the Higgs Particle is of such high priority it was decided to keep the LEP going for a few more months in order to amass more data. No further evidence appeared, and the scientists involved now concede that their claim was premature. However, the consensus remains that Higgs Particles will very soon be generated, quite probably by the Tevatron, a very powerful instrument run by the Fermi National Accelerator Laboratory ('Fermilab') at Batavia, Illinois, USA.

A major aim of theoretical physics is to build a mathematical framework that will explain all the forces that act on matter. In the nineteenth century, the forces of electricity and magnetism were shown to be the same, and were called the 'electromagnetic force'. The Standard Model unites this force with the weak atomic force, which is generally now called the 'electroweak force'. At the close of the twentieth century, theorists were working on various versions of a 'Grand Unified Theory' that would unite the electroweak force with the strong atomic force. Uniting these with the last of nature's forces, gravity, still did not seem possible within the foreseeable future.

The future

It is tempting to close this book with some speculation about what scientific discoveries may be made in the next one hundred years. However, there is a risk of hubris. One only has to think back to the year 1900, and ask: How many of the discoveries described in this book could have been predicted then? Even H. G. Wells, perhaps the most far-sighted prophet that science has ever had, only foresaw a tiny fraction of what was to come. In 1900, nobody would ever have guessed at such discoveries as DNA, the Big Bang, vitamins, or even neutrons. Arthur C. Clarke, in many ways Wells' chief successor, has said that any science sufficiently advanced will be indistinguishable from magic to people who do not understand it. The science of the year 2000 would certainly seem purest magic to the scientists of 1900.

With this caveat in mind, here are just a few predictions for the kind of discoveries that may await us in the next few decades.

Medicine. Knowledge of genetics will lead to tailor-made drugs for individual patients, or at least for individual strains of bacteria and viruses, instead of the general-purpose antibiotics and antiviral drugs of today. However, these will have only limited impact on the general state of peoples' health. The main route to better health for all lies not in scientific discoveries, but in public health programmes, improved diet, sanitation and education. This is especially true in the Third World, but it applies to the industrially developed nations as well.

General Biology. Further discoveries will be made in how organisms regulate their growth. The workings of the brain, and particularly memory, will become better understood.

Genetics. The true function of the seemingly useless 'junk DNA' that takes up so much of our genomes will be discovered. Improved techniques will be devised for manipulating genes, making cloning and gene therapy more practical and reliable.

History of Life. The question of how life began will remain largely unanswered for some time to come. Discoveries will be made about how DNA, RNA and proteins first came into existence, but it will still be hard to see how they came together to make the first living organisms. More fossils of apes dating from 5 to 6 million years ago, when the human and chimpanzee lineages were diverging, will be described. However, none of these discoveries will stop some religious fanatics from denying the fact of evolution, and teaching 'creationism' in its place.

Chemistry. Further discoveries concerning what actually happens on an atomic scale during a chemical reaction will enable chemists to synthesise new substances much more cheaply and with much higher levels of purity than at present. It will also make possible the synthesis of substances in minute quantities—even single molecules with a required architecture in a desired setting. This will lead to a revolution in 'nanotechnology', the building of microscopically small machinery.

Earth and Environmental Sciences. One of the biggest questions is: Will environmental scientists be able accurately to predict the effects of global warming before they actually happen, and it is too late to do anything about them? Great progress has been made in this field in the last twenty years, but there is still a long way to go.

Astronomy. The true nature of the universe's 'missing mass' will be found. We will find out whether the expansion of the universe really is speeding up or slowing down. We will be able to answer finally whether life has ever existed elsewhere in the Solar System.

Physics. The Higgs Particle will be found. The Standard Model will be revised to incorporate it into a Grand Unified Theory, in which the Strong and Weak Atomic Forces, together with electromagnetism, will all be united in one mathematical framework. The task of uniting these forces with the force of gravity lies much further in the future. It will be necessary to develop a quantum version of Einstein's General Relativity theory to describe gravity in a way that is compatible with descriptions of the other forces.

The amount of effort and expenditure that the world has devoted to science has increased immeasurably during the last century. Science will continue to expand, although perhaps not quite so fast. (This depends largely on whether the economic growth of the industrialised nations continues unabated.) For the next decade at least, much of this expansion will take place in genetics and biology, where more and more of the research will be done by commercial companies, rather than the academic laboratories which led the way in the past. Commercial interests may also come to the fore in quantum physics, bringing about yet more revolutions in computer technology.

In 1900, some scientists thought that the end of their quest was in sight. Instead, it was only just beginning. By the year 2100, the science of 2000 may look as quaint as the science of 1900 seems to us today.

Index

Page numbers given in italics refer to entire essays. Numbers given in bold print refer to plates.

Abelson, Philip *192–3*
Absolute Zero temperature 214, 244
Acetyl co-A 80
Acquired Immune Deficiency Syndrome (AIDS) 44–5, 121
Action potential (nervous) *88–9*
Adaptive optics 252
Adenine 83, 106, 112–13, 116
Adenosine diphosphate 82
Adenosine triphosphate *82–3*, 184
ADP, *see* Adenosine diphosphate
Adrian, Edgar 27
Aedes aegypti 18–19
Aerosol cans 180
Afar, Ethiopia 249
Africa 164, 249
Africans 34
Agglutinins 20–1
Agglutinogens 20–1
Agriculture 5, 187
AIDS, *see* Acquired Immune Deficiency Syndrome
Aircraft, supersonic 251
Alaska 176
Alchemy 207
Alcoholism 15
Aldrin, Buzz (Edwin) 9
Aliens, *see* Extraterrestrial life
Alkaptonuria 103
Allen, Willard *78–9*
Alpha-particles 216, 219, 222–3, 230–1, 234
Alpha rhythm 27
'Alternative' medicine 15–16
Aluminium 251–2
Aluminium triethyl 194–5
Alvarez, Luis *60–1*
Alvarez, Walter *60–1*
Alvin (research submarine) 178
Amber 190
Amide group 91
Amino-acids 55, 90–1, 115–17, 124, 157
Ammonia 5, 55, 77, 156–7, 186–7
Anaemia, pernicious 93
Ancestral humans (hominids) 51–3, 62–5

Anderson, Carl David *228–9*
Anderson, M. H. *244–5*
Andes, the 174
Andromeda Nebula xvi, 138–9, **6**
'Animal rights' campaigns 9
Antarctica 166, 175, 250
Antibiotics 15, *28–9,* 106, 111, 256
Antibodies 33, 96, 97, 126–7
Antibodies, monoclonal *96–7*
Anti-particles 229
Anti-protons 243
Anti-quarks, *see* Quarks
Anti-science attitudes 8–9, 15–16
Antiseptics 14
Antiviral drugs 256
Anti-vivisection movement 8, 16
Apollo programme xvi, 147
Apoptosis 248
Appleton, Edward 173
Archaea 179
Archaeopteryx 249
Ardipithecus ramidus 249
Arecibo radio observatory, Puerto Rico 161
Argentina 249
Argon 188
Armstrong, Neil 9
Arnison, G. *242–3*
Arsenic 28
Arthritis, rheumatoid 41
Asteroid impacts 51, 57, 60–1
Asteroids 146
Astronomy, twentieth-century 134–7, 252–4
Astrophysical Journal 139
Astrophysics 135
Athenaeum Club 232
Atlantic Ocean 164, 174
Atmosphere, Earth's 54–5, 77, 172–3, 180–1, 219, 229, 252
Atomic bomb, *see* Nuclear bombs
Atomic nucleus, *see* Nucleus, atomic
Atomic theory 182
Atoms 182–3, 188–9, 200, 205–11, 214–15, 224–5, 232–3, 244–5
ATP, *see* Adenosine triphosphate
Australia 175
Australopithecus afarensis xvi, *62–3*, 249, **10**
Australopithecus africanus *64–5*
Australopithecus species 62–3
Austria 10, 234
Avery, Oswald *106–7*
Axon 88–9

Bacteria 14, 18, 28–31, 40, 43, 47, 58–9, 72–3, 77, 96, 106–7, 110, 126, 177, 248
Bacteriophages 28, *72–3*, 116, 124
Bahamas 176
Baja California 178
Bakelite 190
Ballard, Robert 178
Balloon flights 219
Baltimore, David *120–1*
Banting, Frederick 11, *24–5*
Bardeen, John 215, *236–7*
Barium 199
Barre-Sinoussi, Françoise *44–5*
Baryons 240–1
BASF Laboratories 195
Bassham, James 94
BBC, *see* British Broadcasting Corporation
Beadle, George 102–3
Beans, jack 76
Beatles, The 63
Becker, Herbert 231
Bednorz, Georg *198–9*
Beerbohm, Max 203
Bees, *see* Honey-bee: dances
Behaviour, human 68, 71
Behavioural sciences 66, 84–5

Bell, Jocelyn *154–5*
Bell Telephone Corporation 153, 236
Benson, Andrew 94
Beppo-SAX satellite 159
Berkeley, University of California 94–5, 122, 188, 193
Berger, Hans *26–7*
Berners-Lee, Tim 202
Best, Charles *24–5*
Beta-decay 238
Beta-galactosidase 118–19
Beta-particles 238
Bethe, Hans *144–5*
Bible, The 146
'Big Bang' (cosmology) 135, *152–3*, 229, 239, 253, 256
'Big science' 2–6, 216
Biochemistry 66, 184
Biological Time-Bomb, The (G. R. Taylor) 132
Biology, twentieth-century 66–9, 184
Biotechnology 73
Bismarck (shipwreck) 178
Black-body radiation 204
Black holes 151, 159
'Black smokers' (hydrothermal vents) *178-9*
Blackett, Patrick 229
Blastopore 75
Blink microscope 142
Blood groups 20–1
Blood poisoning 30–1
Blood transfusion 15, 20–1
Bohr, Niels xv, 34, 189, 201, 217, *224–5*, 226, 231, 236–7
Bomb, atomic, *see* Nuclear bombs
Bomb, Hydrogen, *see* Hydrogen bomb
Bone marrow 97
Bose, Satyendra Nath 244
Bose-Einstein Condensate *244–5*
Boston University 59
Bothe, Walter 231
Bovine Spongiform Encephalopathy *42–3*, 115
Bragg, W. L. 93
Brain 26–7, 88, 215, 256
Brattain, Walter *236–7*
Britain, *see* United Kingdom
British Association 233
British Broadcasting Corporation 173
British Empire 1
Brown, Richard 208
Brownian motion 208–9, 220
BSE, *see* Bovine Spongiform Encephalopathy
Buckminsterfullerene 11, 183, 196–7, 251
Burbidge, Geoffrey *148–9*
Burbidge, Margaret *148–9*
Butane gas 189

Cadmium chloride 239
California 175
California Institute of Technology 147, 229
Calvin, Melvin *94–5*
Calvin Cycle *94–5*, 184
Cambrian Period 56
Cambridge University (*see also* Cavendish Laboratory) 23, 27, 49, 88, 90, 112–13, 224, 229, 234, 252
Canada 176
Cancers 41, 122–3, 180, 248
Capitalism 100
Carbohydrates 22–3, 107
Carbon 77, 145, 189, 194–7
Carbon dioxide 55, 80–1, 83, 94–5, 223, 250
Carbon isotopes 57, 94
Carbon-14, radioactive 94
Carlsberg Brewery 225, 236

Carnegie Institute, Washington, DC 162
Carothers, Wallace 191
Carroll, James 18–19
Catastrophism 50, 51
Cavendish Laboratory, Cambridge, UK xv, 223, 232, 234, 2
CD95 death receptors 248
Celera Genomics, Inc. 247
Cells 56–9, 132–3, 248
Celluloid 190
Cellulose 190
Centipedes 128–9
Central America 174
'Central Dogma' (genetics) *114–15*
Cepheid stars 139
Ceramics, synthetic 184
CERN Laboratory, Geneva, Switzerland xvi, 202, 243
CFCs, *see* Chlorofluorocarbons
Chadwick, James *230–1*
Chain reaction, nuclear 192–3, 231, 235
Chaos theory 166
Charing Cross Hospital 71
Charon (moon of Pluto) 143
Chemical bonds *188–9*
Chemical weapons 10, 89, 184, 187
Chemistry, twentieth-century 182–5, 251–2
Chicago, University of 54, 138
Chickens 121–3
Chile 186
Chile saltpetre, *see* Potassium nitrate
Chimpanzee 45, 63, 249, 257
China, Peoples' Republic of 1, 9, 11, 249
Chinese language 11
Chlorella pyrenoidosa 94–5
Chlorine 180–1
Chlorofluorocarbons 180–1
Chloroplasts 59
Chromosomes xvii, 59, *104–5*, 106, 110–11, 121, 128–9
Cigarettes, *see* Smoking
Citric acid 80
Clarke, Arthur C. 136, 256
Cleveland Museum of Natural History, USA 62
Cloning 17, *132–3*, 256
Coal 177
Cobalt 93
Cockroft, John 232, 234
Collip, J. B. 25
Colorado, University of 245
Complementarity 225
'Complementary' medicine, *see* 'Alternative' medicine
Compton, Arthur 211
Computers 93, 258
Conjugation, bacterial ('bacterial sex') 109, 111, 119
Continents 171
Cooper, Leon 215
Cooper Pairs 199, 215
Copenhagen (*see also* Institute of Theoretical Physics) 199
Copper 199, 214
Core, Earth's *168–9*
Cornell, Eric *244–5*
Cornell University, USA 76, 83
Corner, George Washington *78–9*
Corpus luteum 79
Cosmic background radiation 135
Cosmic rays *218–19*, 229, 240
Cosmological Constant 253
Cosmology, *see* Astronomy
Coulson, A. R. *124–5*
Covalent bonding 189
Cowan, Clyde *238–9*
Creation myths 52

Creationism 67–8, 257
Crick, Francis xvii, 3, 34, 67, *112–17*, 120
Croatia 164, 170–1
Crowfoot, Dorothy, *see* Hodgkin, Dorothy
Crust, Earth's 146, 164, 168, 170–1
Cyclotron 193
Cytosine 106, 112–13, 116

Dalton, John 182
Dart, Raymond *64–5*
Darwin, Charles 66, 70, 86–7, 99, 115
Databanks, online 91
Dawkins, Richard 68
Dawson, Charles 52
De Broglie, Victor 211
Deep Elastic Scattering experiment 241
Degenerative diseases 48–9
Denmark 225
Deoxyribonucleic acid (DNA) xvii, 3, 34, 45, 54, 58, 72, 83, 98, *106–7*, 108–11, *112–13*, 114–17, 120–1, 124–5, 130–1, 184, 247, 256–7, **11**, **12**
Descartes, René 226–7
Determinism 226–7
Detroit 199
Deuterium 144, 232–3
D'Herelle, François *72–3*
Diabetes 15, 16, 24–5, 41, 90
Diamond 196
Dinosaurs 50–1, 56, 60–1, 177, 249
Diphtheria 15
Diplococcus pneumoniae 106–7
Dirac, Paul 228–9, 236–7
DNA, *see* Deoxyribonucleic acid
'DNA Fingerprinting' *130–1*
Dogs 70–1
Doll, Sir Richard *36–7*
'Dolly' (cloned sheep), *see* cloning
Domagk, Gerhardt 30–1
Dorsal Lip 75
'Double Helix' DNA molecule *112–13*, 116
Drosophila melanogaster (fruit fly) xvii, 105, 119, 128–9, 248, **14**
Drugs, anti-acid 46
Drugs, psychoactive 89
Du Pont Corporation 191
Duodenum 46
Dyes, synthetic 30, 182
Dysentery, Shiga 72–3

E. coli, see *Escherichia coli*
Earth, age of 57, *146–7*
Earth sciences, twentieth-century 164–7, 250–1
Earthquake vibrations 169
Earthquakes 167–71, 174–5
Earthworms 128
Ecology 66–7
'Eco-warriors' 9
Egg cells 104, 132–3
Ehrlich, Paul 30
Einstein, Albert xv, 141, 173, 201, 205, *208–13*, *220–1*, 228, 235, 244, **1**
Electricity 198–9, 200, 214–15
Electrocardiogram 26
Electroencephalogram *26–7*
Electrolysis 252
Electromagnetic field 210
Electromagnetic force 242, 255, 258
Electron microscopy 67
Electrons 155, 183, 188–9, 199, 200, 210–11, 214–15, 217, 222, 224–6, 228–31, 236–8, 242
Electrophoresis 131
Electroscope 218–19
Electroweak force 255

Elements, chemical 148–9
Embryology 74–5
Enders, John 38
English language 10–11
Enlightenment, Age of 9
Environmental degradation 187
Enzymes 76–7, 102–3
Enzymologia, journal 81
'Epigenesis' doctrine 74–5
Epilepsy 27
Escherichia coli 108–9, 118
Ethics of science 8, 16, 100–1
Ethology 84
Ethylene 194–5
Eugenics 99–100
Eukaryotes 58–9
'Eureka moment' v–vi, 95, 175, 217
Europa (moon of Jupiter) 136, 254
Europe 175, 181, 249
European Union 1
Evans, Martin 49
Evolution theory 66–8, *86–7*, 99, 115, 147
Ewing, John *176–7*
Expanding Universe theory, *see* 'Big Bang'
Explosives 5, 187
Extinctions, mass 50–1, *60–1*, 177
Eyes 129

Faraday, Michael 201
Farman, J. C. *180–1*
Fats 22–3, 107
Feathers 249
Feminism 78
Fermi, Enrico 236–7, 242
Fermi National Accelerator Laboratory, USA 255
Fertilisers, nitrate 5, 186–7
Fire extinguishers 180
First World 1, 9
Fischer, Max 195
Fisher, R. A. 86
Fishes 128, 129
Flagellum 59
Fleming, Alexander xvii, *28–9*
Flood, Noah's 50
Florey, Howard 6, 29
Foams, plastic 180
Foetus, human 48–9
Foot-and-mouth disease 19
Formaldehyde 157
Fossil fuels 177, 250
Fossils 50–2, 56–7, 60, 62–5, 249, 257
Fowler, William *148–9*
Frail, D. A. 161
France 10
Frankenstein 9, 88
Franklin, Rosalind 3, 112–13
Fray, Derek 252
Free will 68
Freedman, Wendy *162–3*
Freemantle Hospital, Australia 46
French language 10
Freud, Sigmund 26
Friedmann, Jerome *240–1*
Frisch, Karl von 67, *84–5*
Fritschi, J. *190–1*
Frogs 132
Frontal lobotomy 8
Fruit Fly, see *Drosophila melanogaster*
Fuller, Buckminster 197
Fullerenes 197, 251
Fungi 129

Galapagos finches 67, 87
Galapagos Islands 87
Galaxies 135, *138–9*, 151–3, 162–3, 253
Galileo 200

Gallo, Robert 44–5
Galvani, Luigi 88
Gamma-ray bursters *158–9*
Gamma-rays 219, 239, 252
Gangrene 28–9, 31
Garrod, Archibald *102–3*
Gastritis 47
Gearhart, John *48–9*
Gehring, Walter 128–9
Geiger, Hans 216–17
Gell-Mann, Murray 240–1
GenBank 247
Gene therapy 256
General Electric Company, Inc. 2
General Relativity, theory of 141, 151, 203, *220–1*, 253, 258
Genes 58–9, 72, 90, 98–100, *102–3*, 104–24, 126–30, 133, 247–8
Genesis, Book of 52
Genetic code *116–17*, 131
Genetic engineering, *see* Genetic modification
Genetic modification 49, 100–1, 117, 121
Genetic mutation 116, 122–3
Genetic sequence *124–5*
Genetically modified foods 8
Genetics 66, 73, 98–133, 184, 256
Geneva observatory 161
Genome 100, 108–9, 110, 121, 124–5
Geological Society of London 169
Geology, *see* Earth sciences
German language 10
Germanium 237
Germany (*see also* Nazi Regime) 3–4, 8, 10, 164, 187, 192
Gestapo 225
Geysers 179
Gibson, T. *32–3*
Glasgow Royal Infirmary, UK 32
Global Positioning System (GPS) satellites 221
Global warming 167, 250, 257
Glucose 95, 118–19
Gluons 242
Glycine 157
Gold 155, 187, 216–17, 222
Gonadal ridge 49
Gorgas, William 19
Graham, E. A. *36–7*
Grand Unified Theory 255, 257–8
Graphite 196
Gravity 220–1, 253, 255, 258
Gray, Tom 62
Green Bank radio observatory, USA 156
'Green revolution' 12
'Greenhouse effect' 250
'Greenhouse gases' 167, 177, 251
Greenland 250
Greenstein, J. L. *150–1*
Greenwich, Royal Observatory 4
Group theory 240
Growth regulation 123
Guanine 106, 112–13, 116
Guano 186
Guinness, Alec 194
Gulf Stream 250

Haber, Fritz 5, 184, *186–7*
Haber Process 5, 184, *186–7*
Hadar (Ethiopia) 62
Haemoglobin 34–5
Hahn, Otto *234–5*
Haile-Selassie, Yohannes 249
Harvard University 191
Havana, Cuba 19
Hawking, Stephen 203
Hazard, C. *150–1*
Heaviside, Oliver 172–3
Heaviside Layer, *see* Ionosphere

Heisenberg, Werner xv, *226–7*, 228
Helicobacter pylori *46–7*
Helium 144–5, 163, 188, 196, 207, 233
Helium, liquid 198, 214–15
Hepatitis 41
Hess, Harry 164–5
Hess, Victor *218–19*
Hewish, Anthony *154–5*
Higgs Field, Higgs Particle 255, 257
Hill, Bradford *36–7*
Himalayas, the 175
Hiroshima 193
History of Life 50–3, 249, 257
Hitler, Adolf 31, 201, 234
HIV, *see* Human Immunodeficiency Virus
Hodgkin, Alan *88–9*
Hodgkin, Dorothy (Crowfoot, Dorothy) *92–3*
'Holes' (quantum) 237
Hollister, Charles *176–7*
Holtzkamp, E. 194
Homeobox, *see* Homeotic genes
Homeotic genes *128–9*
Hominids, fossil (*see also* Ancestral Humans) 64–5
Homo habilis 62
Homogentistic acid 103
Honey-bee: dances xvii–xviii, *84–5*, **15**
Hopkins, Frederick Gowland *22–3*
Hormones 78–9
Hoyle, Sir Fred *148–9*, 152
Hubble, Edwin xvi, 135, *138–41*, 253
Hubble space telescope 162, 252
Hubble's Law 140–1, 150
Human Genome Project xvii, 110–11, 121, 247
Human Immunodeficiency Virus (HIV) *44–5*, 121
Humanism 10
Huxley, Aldous 89
Huxley, Andrew *88–9*
Hybridoma cells 97
Hydrocarbons 194
Hydrogen 55, 77, 144–5, 148–9, 163, 186, 188, 222–4
Hydrogen bomb 201, 232–3
Hydrothermal vents, *see* Black Smokers
Hyena 65

I.G. Farbenindustrie 2, 30–1
IBM research laboratory, Switzerland 198
Ice Ages 250
'Ice dwarfs' 143
ICI Ltd. 194–5
Illinois, University of 215
Immune response 33
Immune system 96–7, 248
Immunosuppressant drugs 33
Incas, *see* Peruvian native Indians
India-rubber 190–1
Indonesia 174
Industrial Revolution 2
Infantile paralysis, *see* Poliomyelitis
Infrared light 252
Insects 128
Institut Pasteur, Paris 44
Institute of Theoretical Physics, Copenhagen 225, 226, 236
Insulin 11, 18, *24–5*, 90–1, 93, 184
Integrated circuit (*see also* Silicon chip) 184, 202
Interferometry 252
Interferons *40–1*
International Geophysical Year 166
Interstellar dust 135, 157
Interstellar gas 135, 156

Ionic bonding 189
Ionosphere *172–3*
Iridium layer 60–1
Iron 149, 169, 187
Isaacs, Alick *40–1*
Isoprene 191
Isotopes, radioactive 146, 192–3

Jacob, François *118–19*
Japan 4, 9, 11, 213
Japanese language 11, 197
Jeffreys, Alec *130–1*
Jena, University of, Germany 27
Jewish culture, liberal 10
Jewish scientists 10, 187, 201, 225, 227, 234–5
Johanson, Donald xvi, *62–3*
Johns Hopkins University, Baltimore, USA 48
Johnson, W. A. *80–1*
Journal of Geophysical Research 175
Jung, Karl Gustav 26
'Junk DNA', *see* Non-functional DNA
Jupiter 136, 143, 157, 160–1

Kaiser Wilhelm Institute, Berlin 74, 234
Kaufman, M. H. 49
Kendall, Henry *240–1*
Kennelly, Arthur 172
Ketterle, Wolfgang 245
King's College, London University 3, 112, 173
Kohler, Georges *96–7*
Krebs, Hans *80–1*
Krebs Cycle *80–1*, 83, 184
Kroll Process 252
Kroto, Sir Harry *196–7*
Krypton 188
Kupa Valley 170

Lack, David 67, *86–7*
Lacquer 190
Lactose 118–19
Landsteiner, Karl *20–1*
Langerhans, Islets of 24–5
Lanthanum 199
Large Electron-Positron Collider (LEP) xvi, 255, **5**
Lasers 196, 245
Lazear, Jesse 18–19
Lead isotopes 146–7
Lederberg, Joshua *108–9*
Lehmann, Inge 169
Leicester University 131
Lentiviruses 45
Leopard 65
Leukaemia 41
Leukaemia viruses 44–5
Lewis, Gilbert 34, *188–9*
Life, extraterrestrial 136–7, 154, 157, 253–4, 257
Life, origins of 51, *54–5*, 179
Lifespan, increase in 14–15
Light 201–5, 208, 210–13, 224, 242
Lindemann, J. *40–1*
Lipmann, Fritz *82–3*
Lithium 163, 232–4
Lobsters 128, 129
Lohmann, Kurt 83
Lorenz, Konrad 85
Los Alamos National Laboratory, USA 238–9
Lowell, Percival 142–3
Lowell Observatory, Arizona, USA xvi, 142
'Lucy' (pre-human ancestor), see *Australopithecus afarensis*
Lung cancer 36–7
Lungs 34–5
Lymphocytes 126–7
Lymphocytes, B 96–7
Lymphocytes, T 44–5
Lysenko, Trofim 100

Macleod, J.J.R. 25
'Mad Cow Disease', *see* Bovine Spongiform Encephalopathy
Magnesium 252
Magnetic field, Earth's 166
Magnetic Resonance Imaging, *see* MRI Scanner
Maize 110–11
Malaria 28, 35
Mammary gland 132–3
Mammoth 56
Man in the White Suit, The (film) 194
Manchester, University of 3, 216, 222, 224
Mangold, Hilde, *see* Proescholdt, Hilde
Mangold, Otto 75
Manhattan Project 3–4, 235
Mantle, Earth's 164–5, 171
Marconi, Guglielmo 172
Margarine 23
Margulis, Lynne, *see* Sagan, Lynne
Mars 136, 178, 254
Marsden, Ernest 216–17, 222
Marshall, Barry *46–7*
Martin, G. Steven *122–3*
Martin, Gail 49
Marxism, Marxist-Leninism 1, 71, 100
Mass 212–13, 221, 238–9, 253
Massachusetts Institute of Technology 120, 126, 245
Matrix Mechanics theory 226
Matthews, H. C. 27
Matthews, T. A. *150–1*
Max Planck Institut, Mulheim, Germany 194
Maynard Smith, John 67
Mayor, Michel *160–1*
McClintock, Barbara *110–11*
McGill University, Canada xv, 206–7, 216, **4**
McGinnis, William *128–9*
McMillan, Edwin *192–3*
Medawar, Sir Peter *32–3*
Medical Research Council, UK 5, 36
Medicine, twentieth-century 14–17, 184
Mediterranean Sea 174
Meitner, Lise 234–5
Mendel, Gregor 98, 102, 104–5
Mendeleev, Dmitry 182
Mercury (metal) 214
Mercury (planet) 220–1
Mesons 240–1
Metals 189
Meteorites 146–7, 254
Meteorology, *see* Earth sciences
Methane clathrate *176–7*
Methane gas 54–5, 176–7
Mexico 178
Mice 97
Microwave radiation, cosmic 153
Microwave radio transmission 153
Military research 166
Milky Way Galaxy 135–6, 138, 157, 160
Mill Hill, *see* National Institute of Medical Research
Miller, Stanley *54–5*
Milliken, Robert 211
Milne, John 168, 171
Milstein, Cesar *96–7*
Minisatellites (DNA) 131
'Missing mass' (cosmology) 253, 257
Mitochondria 59
Mitsuzani, S. *120–1*
Moho, the (Mohorovicic Discontinuity) *170–1*
Mohorovicic, Andrija 164, 168, *170–1*
Mojzis, S. J. *56–7*

Molecular structures *92–3*
Mongolia 249
Monod, Jacques *118–19*
Montagnier, Luc 44–5
Montreal Protocol 181
Moon, The 147, 158, 178
Morgan, Thomas Hunt xvii, 99, *104–5*
Morgan, W.J. xvi, 175
Mosquito, Yellow-fever, see *Aedes aegypti*
Mount Palomar Observatory, California 150, 153, 159
Mount Wilson Observatory, California 135, 138
Mountain ranges 174–5
MRC, *see* Medical Research Council
MRC Molecular Biology Laboratory, UK 97, 124
MRI scanner 215
Muller, Alex *198–9*
Multiple sclerosis 41
Munich, University of 245
Munich Botanical Garden 84
Munster, University of, Germany 30
Muslim culture 10
Mutant genes 103–4
Myeloma cells 97

Nagasaki 193
Nanotechnology 257
Nanotubes 197, 251
National Cancer Institute, USA 44
National Foundation for Infantile Paralysis, USA 39
National Institute of Medical Research, UK 40
National Institutes of Health, USA 247
Natta, Giulio 191, 195
Natural Gas 176–7
Nature (journal) 81, 120, 150
'Nature versus nurture' 68
Nazi Regime (*see also* Germany) 4, 8, 27, 99, 201, 213, 225, 227, 235
Nebulae 134, 138, 140
Neo-Darwinism 99
Neon 188
Neptune 142–3
Neptunium 193
Nerve gases, *see* Chemical weapons
Nervous system 88, 248
Neuroanatomy 26
Neurology 89
Neutrinos *238–9*
Neutron stars 155
Neutrons 144, 148–9, 155, 183, 193, *230–1*, 234–5, 238, 240–2, 256
New York Times 139
Newt, Alpine 74–5
Newt, Crested 74–5
Newton, Sir Isaac 200–1, 220, 226
Nickel 194–5
Nitrogen 55, 77, 145, 186–7, 223
Nitrogen, liquid 199
Nitrosoguanidine 122
Nobel Prize 10–11, 25, 31, 33, 38, 67, 70, 85, 90, 93, 103, 125, 149, 189, 195, 210, 215, 232, 237
Non-functional DNA ('junk DNA') 130–1, 256
North America 174–5, 181, 249
Nuclear bombs (*see also* Manhattan Project) 4, 8, 145, 158, 192–3, 201, 213, 225, 227, 231, 234–5
Nuclear disarmament 201, 213
Nuclear energy 193, 233
Nuclear fission *234–5*

Nuclear fusion 144–5, 150, 196, *232–3*
Nuclear pile 192–3, 227, 231, 234–5, 239
Nuclear reactor, *see* Nuclear pile
Nuclear war 93
Nucleic acids 54, 90, 107
Nucleotides 55
Nucleus, atomic *216–17*, 222–5, 230–5, 241
Nucleus (in living cells) 121
Nylon 191

Occhialini, Giuseppe 229
Oceans 171, 174–5, 177
Oestrogen 78
Office of Scientific Integrity, USA 45
Oil 165, 177
Oke, J. *150–1*
Oldham, Richard *168–9*
Omega Minus particle 240
Oncogenes *122–3*
Onnes, H. Kammelingh *214–15*
Oparin, Aleksander 51
Operator genes 119
Operon *118–19*
Opie, Eugene 24
Orang-utan 52
Organic molecules 156–7
Osawa, E. 197
Osmium 186–7
Ovaries (in mammals) 78–9
Oxaloacetic acid 80
Oxford University 6, 28, 32, 92, 138, 207
Oxygen 54, 77, 145, 223
Ozone loss, atmospheric *180–1*

Pacific Ocean 174–5, 178
Pacifism 93, 201, 213, 225
Palaeontology, twentieth-century, *see* History of Life
Panama Canal 19
Pancreas 24–5
Parkes radio observatory, Australia 150
Parkinson's Disease 48, 89
Particle accelerators 202
Particles, sub-atomic, *see* Sub-atomic particles
Pasteur, Louis 18
Patent medicines 6, 14
Patterson, Claire *146–7*
Paulesco, Nicolas 11, 25
Pauli, Wolfgang 238–9
Pauling, Linus *34–5*, 189, 225
Pavlov, Ivan Petrovich *70–1*, 84
Pea plants 102, 119
Penicillin xvii, 6, *28–9*, 30, 92, **13**
Penicillium notatum xvii, 29, **13**
Pentagon, The 158
Penzias, Arno *152–3*
Periodic Table of chemical elements 182
'Permissive society' 78
Perovskites 199
Persian Gulf 165
Peruvian native Indians 20–1
Phages, *see* Bacteriophages
Pharmaceutical companies 6–7, 247
Pharmaceutics 184
Phenylalanine 103
Phosphate ion 82, 106, 112
Phosphoglyceric acid 95
Photoelectric effect 173, 210–11
Photography 134
Photons *210–11*, 220
Photosynthesis 94–5
Physical Review, The (journal) 237
Physics 135, 183, 257–8
Physics, twentieth-century (modern) 200–3, 254–5
Physiology 66–7
Pill, contraceptive 8, 78–9

Piltdown Skull 52
Pincus, Gregory 79
Pitchblende 206
Pittsburgh University, USA 38
Planck, Max 200–1, *204–5*
Planck's Constant 205, 211
Planets 134–5, 142–3, 157, 160–1, 252
Plankton, fossil 60
Plants 56, 59–60, 94–5, 129
Plasmids, genetic 109, 111
Plastics 182, 184, 190–1, 194–5
Plate Tectonics, theory of xvi, 146, 165, *174–5*, 178, **8**
Platinum 155
Pluto xvi, *142–3*, **9**
Plutonium *192–3*
Pneumonia 15, 31, 106
Poland 73
Polio vaccine *38–9*
Poliomyelitis (polio, infantile paralysis) 15, 38–9
Politics, left-wing 68–9
Politics, right-wing 68
Pollution 185
Polonium 206
Polyethylene, *see* Polythene
Polymers *190–1*
Polypropylene 195
Polystyrene 191
Polythene 194–5
Polyvinyl chloride 190
Popular science writing 9
Positrons *228–9*
Potassium ions 89
Potassium nitrate 186
'Preformation' doctrine 74–5
Pregnancy 78–9
Princeton University 175, 193
Prions *42–3*, 115
Proescholdt, Hilde (Mangold, Hilde) *74–5*
Progesterone *78–9*
Prokaryotes 58–9
Prontosil rubrum 30–1
Proteins 22–3, 34, 43, 54–5, 72, 77, *90–1*, 97, 106–7, 110, 114–17, 119, 121, 124, 126, 257
Protestantism 10
Proton-Antiproton Collider 243
Protons 144, 148–9, 155, 183, 211, 219, *222–3*, 230–4, 238–43
Prusiner, Stanley *42–3*
Pseudo-science 7–8
Psychoanalysis 26
Public health programmes 15, 256
Pugwash conferences 93
Pulsars 135, *154–5*, 161
PVC, *see* Polyvinyl chloride

Quantum computer 245
Quantum electrodynamics 242
Quantum mechanical theory 227–8, 236–7
Quantum theory 200, *204–5*, 210, 217, 223–5
Quarks 219, 231, *240–1*
Quasars 135, *150–1*, 156, 253
Quasi-stellar radio sources, *see* Quasars
Queroz, Didier *160–1*
Quinine 28

Rabbits 79
Radio Astronomy 135
'Radio stars', *see* Quasars
Radio telescopes 150
Radio transmission 172–3
Radio waves 156–7, 165–6, 172–3, 242
Radioactive decay xv, 146–7, 200, *206–7*, 218, 222, 238
Radium 206, 222
Rayon 190
Recessive genes 35, 103–5
Recombination, genetic *108–9*

Red blood cells 34–5
'Red giant' stars 148–9
Red-shift 150
Reductionism 69
Reed, Walter *18–19*
Reflexes, conditioned *70–1*
Refrigerators 180
Reines, Frederick *238–9*
Relativity theory, *see* General Relativity; Special Relativity
Religion 51, 68, 136, 147, 257
Renaissance 9
Rennin 76
Repressor genes 119
Resistance, Electrical 214–15
Restriction endonucleases 125
Retroviruses 45, 121
Reverse Transcriptase 115, *120–1*
Ribonucleic Acids (RNA) 45, 54, 115, 120–1, 257
Ribose 106, 112
Ribosomes 115
Ribulose diphosphate 95
Rice University, Houston, USA 196
RNA, *see* Ribonucleic Acids
Rocky Mountains 174
Roman Catholicism 8, 10
Romania 11
Roosevelt, President F. D. 31, 213, 235
Rous, Peyton 120
Rous Sarcoma Virus 120–3
Royal College of Physicians, UK 103
Royal Society, The, UK 23, 232
Rubbia, Carlo *242–3*
Rubidium 245
Russia (*see also* Soviet Union) 4, 8, 10, 71
Russian language 10–11
Rutherford, Ernest xv, 3, 189, 202, *206–7, 216–17, 222–3,* 224–5, 230, *232–3,* 234, 241, **2**
Sabin, Albert 39
Sabre-toothed tiger 56
Sagan, Lynne (Margulis, Lynne) *58–9*
Salk, Jonas *38–9*
Salk Institute, California 39
Salt, *see* sodium chloride
Salvarsan 6, 30
San Francisco, University of California 49
Sanger, Sir Frederick *90–1, 124–5,* 247
Satellites 153, 162, 173
Saturn 143
Sauropods (dinosaurs) 249
Schmidt, Harrison xvi, **7**
Schmidt, M. *150–1*
Schrieffer 215
Schrödinger, Erwin 226, 228
Science fiction 134
Scrapie 42–3
Scripps Institute of Oceanography, USA 56–7
Searle, G. D., Ltd. 79
Segmentation 128–9
Seismology 168–71
Selfish Gene, The (Richard Dawkins) 68
Semiconductors 197, 225, 237
Sex, bacterial, *see* Conjugation, bacterial
Shaw, George Bernard 70
Sheep 132–3
Sheffield University, UK 80
Shelley, Mary 88
Shockley, William 236–7
Siberia, Outer 176
Sickle Cell Anaemia *34–5*
Silicon 237
Silicon chip (*see also* Integrated circuit) 184, 201–2, 215, 236, 245

Slipher, Vesto 142
Smith, John Maynard, *see* Maynard Smith, John
Smoking, link with lung cancer 15, *36–7*
Snow, C. P. 8, 234
'Snowball Earth' 250–1
Snyder, Lewis *156–7*
Society of Experimental Biology 114
Sociobiology 68–9
Sociobiology (E.O. Wilson) 68
Soddy, Frederick xv, *206–7*
Sodium chloride 189
Sodium ions 89
Solar System xvi, 134, 136, 143, *146–7*, 200, 253
Solzhenytsin, Alexander 202
South America 164, 174–5, 249
Soviet Academy of Sciences 100
Soviet Union (*see also* Russia) 4–5, 8, 11, 71, 73, 100, 158, 166
Space-flight 134
Space race 5
Special Relativity, theory of xv, 208, *212–13*, 220, 228, 243, **3**
Spectrograph 135
Spectroscopy 140
Spemann, Hans *74–5*
Sperm cells 104
Spiess, F. N. *178–9*
Spin (in sub-atomic particles) 199, 238–9
Spleen 34
Sputnik 1 satellite 166
Squid 89
St. Mary's Hospital, London 28
Stalin, Josef 100
Stalinism 8
'Standard Model' (physics) 242–3, 254, 257
Stanford University, USA 241
Staphylococcus aureus xvii, 28–30, **13**
Stars 134–6, 138–9, 144–5, 148–9, 151, 154–5, 159, 239
State funding of science 6, 135
Staudinger, Hermann *190–1*
'Steady State' theory (cosmology) 148–9, 152
Stem cells, embryonic *48–9*
Stevenson, George 2
Stomach, human 46–7
Strassmann, Fritz *234–5*
Streptococci, haemolytic 30–1
Strong atomic force 242, 255, 258
Sub-atomic particles (*see also* baryons, electrons, gluons, mesons, neutrinos, neutrons, Omega Minus particle, photons, protons, quarks, W particles, Z particles) 219, 228, 238, 240, 242–4, 254
Submarines, detection of 222
Sulphanilamide 31
Sulphonamides 15, 28, *30–1*, 84
Sumner, James *76–7*
Superconductors, 'high-temperature' 183, *198–9*, 215, 251
Superconductors, low temperature 199, *214–15*, 251
Supernovae 145, 149, 155, 161, 196, 219
Supernovae, Type 1a 163
Sussex, University of 196
Surrogate mother (sheep) 133
Sweden 235
Switzerland 10, 187
Syphilis 6, 28
Szilard, Leo 235

T Lymphocytes 44–5
Taieb, M. *62–3*
Tatum, Edward 102–3, *108–9*

'Taung Child', see *Australopithecus africanus*
Taylor, Richard *240–1*
Temin, Howard *120–1*
Tevatron (particle collider) 255
Things to Come (film) 9
Third World 7, 9, 11, 256
Thorium 146–7, 206–7
Thoron 207
Thymine 106, 112–13, 116
Tinbergen, Nikolaas 85
Tissue culture 67
Titanic (shipwreck) 178
Titanium 251
Titanium chloride 195
Tobacco industry 37
Tobacco smoke (*see also* Smoking) 123
Tombaugh, Clyde xvi, 142–3
Tonegawa, Susumu *126–7*
Toronto, University of 24
Transistor 202, 215, 225, *236–7*
Transplant rejection *32–3*, 49, 96
Transplant surgery 32
Transposons 110–11
Trans-uranium elements 192
Trilobites 56
Turner, Louis 193
'Two Cultures' 8
Twort, F. W. 72
Tyrannosaurus rex 249
Tyrosine 103

UFOs, *see* Unidentified Flying Objects
Ulcers, gastric 46–7
Ultraviolet light 122, 180, 252
Uncertainty Principle *226–7*
Unidentified Flying Objects (UFOs) 134
United Kingdom 8, 10, 187, 192
United States of America 1, 3, 5, 8, 18, 34, 68, 145, 166, 192, 213
Universe, age of 145, 149, *162–3*
Universe, expansion of *140–1*
University, *see under* place names
Uranium 146–7, 149, 192–3, 206, 231, 234–5
Uranus 142–3
Urea 77
Urease *76–7*, 184
Urey, Harold 54–5
Urine 103
Ussher, Archbishop 146
USSR, *see* Soviet Union

Vaccination 15
Valency 182
Valve (electronic) 237
Van der Meer, Simon *242–3*
'Vela' satellite 158
Vening-Meinesz, Felix 165
Venter, Craig 247
Vertebrate Organiser *74–5*
Vienna, University of 20
Virtual particles 243
Viruses 18, 33, 40–1, 43, 72–3, 110–11, 126, 248
Vitamin A 23
Vitamin B12 *92–3*, 184
Vitamin D 23
Vitamins *22–3*, 256
Volcanic eruptions 61, 167, 250
Volcanoes 174
Vulcanite 190

W particles *242–3*
Wallace, Alfred Russell 66, 86
Walton, Ernest 232, 234
War, Cold 158, 166
War, First World 5, 21, 28, 165, 187, 222
War, Second World 3–6, 8, 29, 32, 35, 87, 89, 135, 145, 201, 213, 231, 237
Washing powder, biological 76

Water 156–7, 223
Watson, James xvii, 3, 34, *112–13*, 116, 120, 247, **12**
Wave Mechanics theory 226
Waves (in quantum physics) 210–11, 224, 244
Weak atomic force 242–3, 255, 258
Weather forecasting 165–6
Wegener, Alfred 164–5
Wells, H. G. 9, 256
White, Tim 249
'Whole-organism' biology 67
Wieman, Carl *244–5*
Wilkins, Maurice 112–13
Wilmut, Ian *132–3*
Wilson, Charles 218
Wilson, E. O. 68
Wilson, Robert *152–3*
Wisconsin, University of 120
Witwatersrand University, South Africa 65
Wolzsczan, A. 161
Woods Hole Oceanographic Institute, USA 178
World Wide Web 202
Wynder, E. L. *36–7*

X chromosome 103–4
Xenon 188
X-ray diffraction analysis 67, 92–3, 98, 112

Y chromosome 103–4
Yellow Fever Virus *18–19*
Yoshida, J. 197
Yucatan, Mexico 61

Z particles *242–3*
Zagreb, Meteorological Observatory at 170
Zero point energy 214, 244
Ziegler, Karl 191, *194–5*
Ziegler-Natta catalysts 183, 191, *194–5*
Zurich Patent Office 207, 212, 220
Zurich Polytechnik 190
Zweig, George 240–1
Zymase 76